Weigmann/Kilian   Decentralization with PROFIBUS DP/DPV1

Decentralization with PROFIBUS DP/DPV1

Architecture and Fundamentals,
Configuration and Use with SIMATIC S7

by Josef Weigmann and Gerhard Kilian

2nd revised and enlarged edition, 2003

Publicis Corporate Publishing

Bibliographic information published by Die Deutsche Bibliothek

Die Deutsche Bibliothek lists this publication in the Deutsche Nationalbibliografie; detailed bibliographic data is available in the Internet at http://dnb.ddb.de

ISBN 3-89578-218-1

Editor: Siemens Aktiengesellschaft, Berlin and Munich
Publisher: Publicis Corporate Publishing, Erlangen

Printed in Germany

Preface

In the last decade, rapid progress in the decentralized architecture of automated industrial plants and processes has made fieldbus systems increasingly popular. The reason for this development is obvious: installing I/O channels where they are actually needed – in the vicinity of the machine – reduces installation and wiring work to a minimum. Significant cost savings are the result.

Standardized fieldbus systems with "open" communications interfaces enable the use of distributed inputs/outputs and also of intelligent, process signal-conditioning field devices from different manufacturers.

The architecture of a fieldbus system must be both transparent and open. These are important aspects for automation engineers when they have to decide which fieldbus to choose from among the wide range of components now available on the market.

With its bus access procedure, PROFIBUS meets the important demand for data accessibility on the field level of the industrial plant. On the one hand it covers the communication requirements in the sensor/actuator area, and on the other hand all networking functions for the so-called cell area. Particularly in the area of "distributed I/O," PROFIBUS has become a de facto internationally accepted standard with a large selection of connectable field devices.

This book describes PROFIBUS (PROcessFIeldBUS), an open fieldbus system that complies with the EN (EuropaNorm) standard, Volume 2 [1]/IEC 61158 [10], and whose protocol has been specialized for decentralized peripherals (DP). We have written this book with the goal of making it as easy as possible for system engineers, programmers and commissioning engineers to familiarize themselves with the subject and implement automation tasks with PROFIBUS DP. The use of PROFIBUS DP/DPV1 is illustrated by many example projects based on the SIMATIC S7 automation system and STEP 7 Version 5.1 SP3.

In this second, revised, and updated edition, new interrupts and user program interfaces in the SIMATIC S7 are introduced with regard to the DPV1 expansion.

While it is advisable that you have a basic knowledge of the SIMATIC S7-300 and SIMATIC S7-400 programmable controllers and the STL (statement list) programming language, this is not an essential requirement.

Erlangen, July 2003   Josef Weigmann, Gerhard Kilian

Contents

1 PROFIBUS Fundamentals

Introduction

If we compare an automated plant whose communication is based on a serial fieldbus system with an automated plant installed in the conventional way, you will notice the advantages at first glance. Using industrial fieldbus technology, considerable savings can be made particularly in the mechanical installation, fitting and wiring of the plant equipment due to reduced cabling for distributed input/output devices. A second convincing factor is the wide variety of field devices that are available for this technology. However, to make the most of these advantages the fieldbus must be of standardized design and open architecture. In 1987, German industry has therefore initiated the PROFIBUS Cooperative Project. The regulations and norms developed by this body were documented in the DIN E 19 245 [2] PROFIBUS standards. In 1996, this national fieldbus standard became the international EN 50170 standard.

1.1 ISO/OSI Model

PROFIBUS makes use of already existing national and international standards. The protocol is based on the OSI (*O*pen *S*ystems *I*nterconnection) reference model in accordance with the internal ISO standard (*I*nternational *S*tandard *O*rganisation).

Figure 1.1 illustrates the ISO/OSI model for communication standards.

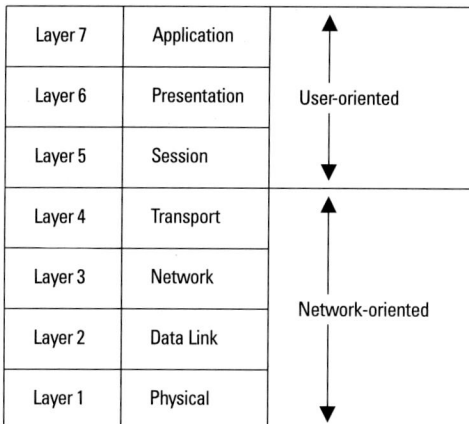

Layer 7	Application	
Layer 6	Presentation	User-oriented
Layer 5	Session	
Layer 4	Transport	
Layer 3	Network	
Layer 2	Data Link	Network-oriented
Layer 1	Physical	

Figure 1.1
ISO/OSI model for communication standards

The ISO/OSI model for communication standards consists of 7 layers and is organized into two classes. One class comprises the user-oriented Layers 5 to 7, the other contains the network-oriented Layers 1 to 4. Layers 1 to 4 describe the transport of data from one location to another, whereas Layers 5 to 7 provide the user with access to the network system in an appropriate form.

1.2 Architecture and Versions of the Protocol

The overview in figure 1.2 illustrates which layers of the ISO/OSI model are implemented for the PROFIBUS protocol: Layers 1 and 2 and, if necessary, Layer 7. The line and transmission protocols of Layers 1 and 2 comply with the USA standard EIA (*E*lectronic *I*ndustries *A*ssociation) RS 485 [8], the international standard IEC 870-5-1 [3] (Telecontrol Equipment and Systems) and the European EN 60 870-5-1 [4] standard. The bus access procedure and the data transmission and management services are based on the DIN 19 241 [5] standard, Parts 1 to 3, and on the IEC 955 [6] standard (Process Data Highway/Type C). The management functions (FMA7) use the ISO DIS 7498-4 (Management Framework) concept.

From the user's point of view, PROFIBUS provides three different versions of the communication protocol: DP, FMS and PA.

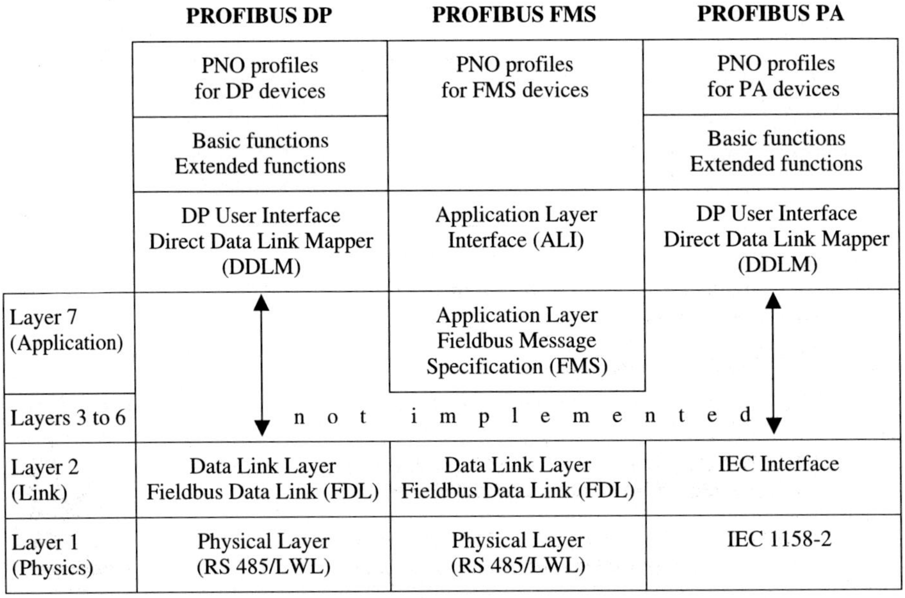

	PROFIBUS DP	PROFIBUS FMS	PROFIBUS PA
	PNO profiles for DP devices	PNO profiles for FMS devices	PNO profiles for PA devices
	Basic functions Extended functions		Basic functions Extended functions
	DP User Interface Direct Data Link Mapper (DDLM)	Application Layer Interface (ALI)	DP User Interface Direct Data Link Mapper (DDLM)
Layer 7 (Application)	↑	Application Layer Fieldbus Message Specification (FMS)	↑
Layers 3 to 6	↓ n o t i m p l e m e n t e d ↓		
Layer 2 (Link)	Data Link Layer Fieldbus Data Link (FDL)	Data Link Layer Fieldbus Data Link (FDL)	IEC Interface
Layer 1 (Physics)	Physical Layer (RS 485/LWL)	Physical Layer (RS 485/LWL)	IEC 1158-2

Figure 1.2   Protocol architecture of PROFIBUS

1.2.1 PROFIBUS DP

PROFIBUS DP uses Layer 1, Layer 2, and the User Interface. Layers 3 to 7 are not developed. This lean architecture ensures high-speed data transmission. The Direct Data Link Mapper (DDLM) provides access to Layer 2. The available application functions and the system and device characteristics of the various types of PROFIBUS DP devices are specified in the User Interface.

Optimized for high-speed transmission of user data, this PROFIBUS protocol is designed especially for communication between the programmable controller and the distributed I/O devices at the field level.

1.2.2 PROFIBUS FMS

In *PROFIBUS FMS*, Layers 1, 2 and 7 are implemented. The Application Layer consists of FMS (*Fieldbus Message Specification*) and LLI (*Lower Layer Interface*). FMS contains the application protocol and provides the communication services. LLI establishes the various communication relationships and provides FMS with device-independent access to Layer 2.

FMS handles data communication at the cell level (PLC and PC). The powerful FMS services can be used in a wide range of applications and offer great flexibility when solving complex communication tasks.

PROFIBUS DP and PROFIBUS FMS use the same transmission technology and bus access protocol. They can therefore run simultaneously on the same cable.

1.2.3 PROFIBUS PA

PROFIBUS PA uses the expanded PROFIBUS DP protocol for data transmission. In addition, it implements the PA profile which specifies the characteristics of the field devices. The transmission technique in accordance with the IEC 1158-2 [7] standard ensures intrinsic safety and powers the field devices over the bus. PROFIBUS PA devices can easily be integrated in PROFIBUS DP networks by using segment couplers.

PROFIBUS PA is especially designed for the high-speed and reliable communication required in automated process engineering. With PROFIBUS PA you can link sensors and actuators to a common fieldbus line, even in potentially explosive areas.

1.3 PROFIBUS Layer

1.3.1 Physical Layer (Layer 1) for DP/FMS (RS 485)

In its basic version for shielded and twisted pair cables, Layer 1 of PROFIBUS implements symmetrical data transmission in accordance with the EIA RS 485 [8] standard (also known as H2). The bus line of a bus segment is a shielded, twisted pair cable which is terminated on both ends (see figure 1.3). Transmission speeds of 9.6 kbit/s to 12 Mbit/s can be selected. The selected baud rate applies to all devices connected to the bus (segment).

Transmission procedure

The RS 485 transmission procedure used for PROFIBUS is based on semi-duplex, asynchronous, gap-free synchronization. Data is transmitted in an 11-bit character frame (see figure 1.4) in NRZ code (*Non Return to Zero*). The shape of the signal during the transition from binary "0" to "1" does not change while the bits are being transmitted.

During the transmission, a binary "1" corresponds to a positive level on line RxD/TxD-P (*Receive/Transmit-Data-P*) as opposed to RxD/TxD-N (*Receive/Transmit-Data-N*). The idle state between the individual telegrams corresponds to a binary "1" signal (see figure 1.5). In specialized literature, the two PROFIBUS data lines are also frequently referred to as the A line and the B line. The A line corresponds to the RxD/TxD-N signal whereas the B line corresponds to the RxD/TxD-P signal.

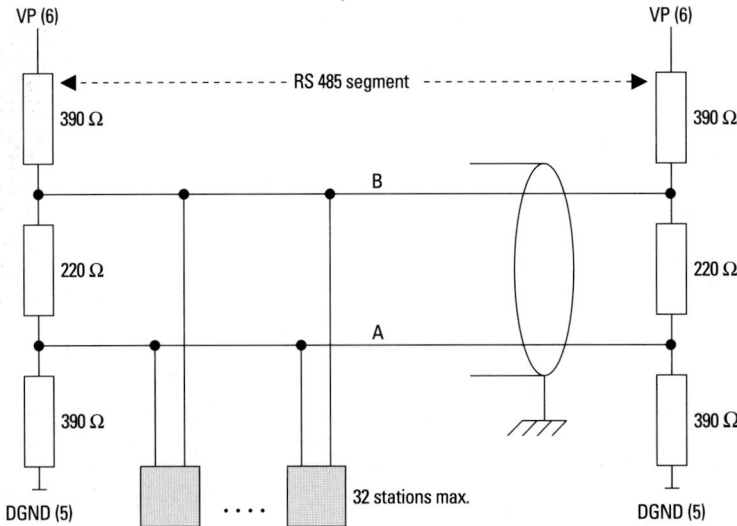

Figure 1.3   Setup of an RS 485 bus segment

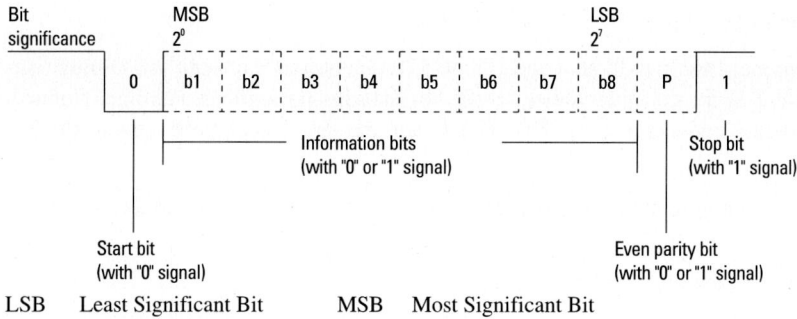

Figure 1.4   PROFIBUS UART character frame

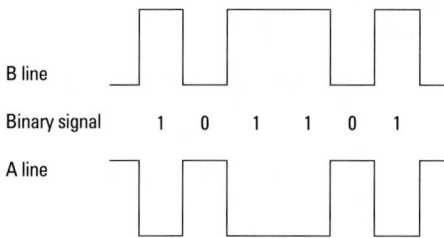

Figure 1.5
Signal form during NRZ transmission

Bus line

The maximum permissible line length, also referred to as segment length, for a PROFIBUS system depends on the selected transmission speed (see table 1.1). A maximum of thirty-two stations can be operated in one segment.

Table 1.1   Maximum segment length based on the baud rate

Baud rate (kbit/s)	9.6 to 187.5	500	1,500	12,000
Segment length (m)	1,000	400	200	100

The maximum segment length specified in table 1.1 refers to cable type A as laid down in the PROFIBUS standard and in table 1.2. This type of cable (A) has the following characteristics.

Table 1.2   Specification of PROFIBUS RS 485 cable, type A

Surge impedance	135 to 165 Ω, at a measuring frequency of 3 to 20 MHz
Cable capacitance	< 30 pF per meter
Core cross section	> 0.34 mm², acc. to AWG 22
Cable type	Twisted pair, 1×2 or 2×2 or 1×4 conductors
Loop resistance	$< 110 \ \Omega$ per km
Signal attenuation	9 dB max. over the entire length of the cable section
Shielding	Braided copper shield or braided shield and foil shield

Bus connection

The international PROFIBUS standard EN 50 170 recommends a 9-pin sub D plug connector (table 1.3) for the interconnection of bus stations through the bus line. Connect the sub D socket connector to the bus station, and the sub D plug connector to the bus cable.

The signals shown in bold type are mandatory signals; they must be available.

Table 1.3   Pin assignment of the 9-pin sub D plug connector

View	Pin No.	Signal Name	Designation
	1	SHIELD	Shield or function ground
	2	M24	Ground of the 24 V output voltage (auxiliary power)
	3	**R×D/T×D-P**	**Receiving/sending-data-plus B line**
	4	CNTR-P	Signal for direction control P
	5	**DGND**	**Data reference potential (ground)**
	6	**VP**	**Supply voltage-Plus**
	7	P24	24 V Plus of the output voltage (auxiliary power)
	8	**R×D/T×D-N**	**Receiving/sending-data-Minus A line**
	9	CNTR-N	Signal for direction control N

Bus termination

In addition to the two-sided bus line termination of data lines A and B of the EIA RS 485 standard, the PROFIBUS line termination includes a pull-down resistor against DGND data reference potential, and a pull-up resistor against the supply voltage-plus VP (see figure 1.3). These two resistors ensure a defined idle potential on the bus line when no station is sending, that is, when the bus line is in the idle state between two telegrams. The required combinations of bus line termination are available on almost all standard PROFIBUS bus connector plugs and can be activated by means of jumpers or switches.

If the bus system runs at transmission speeds higher than 1,500 kbit/s, bus connection plugs with additional longitudinal inductance must be used due to the capacitive load of the connected stations and the resulting line reflections (see figure 1.6).

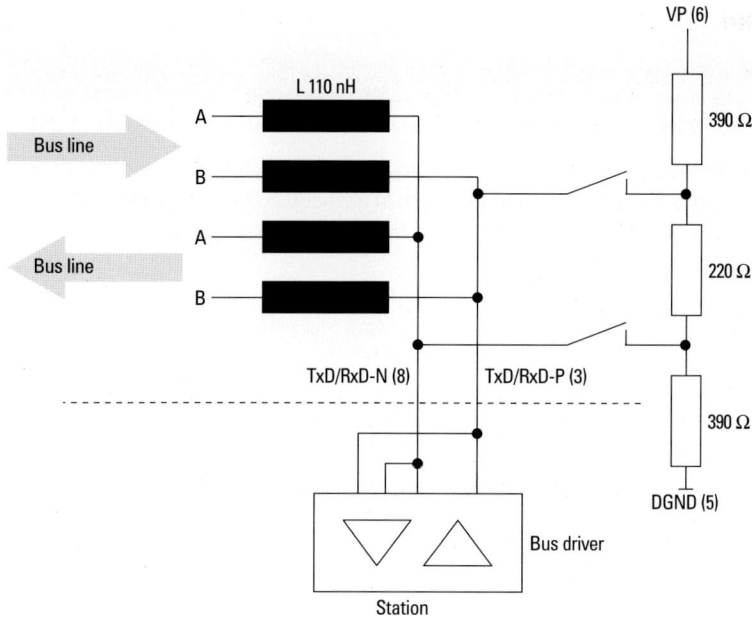

Figure 1.6
Layout of the bus plug connector and bus termination for transmission speeds > 1500 kbit/s

1.3.2 Physical Layer (Layer 1) for DP/FMS (Fiber Optic Cable)

Another version of PROFIBUS Layer 1, based on the guidelines of the PNO (*Profibus Nutzer Organisation* – PROFIBUS International), "Optical Transmission Technology for PROFIBUS, version 1.1 dated 07.1993" [9], is the transfer of data through the transmission of light in fiber optic conductors. Fiber optic cables permit transmission distances of up to 15 km between the stations of a PROFIBUS system. They are not sensitive to electromagnetic interference and always ensure galvanic isolation between the individual bus stations. As the connection technique used for fiber optics has been greatly simplified during recent years, this transmission technology has become very popular for data communication with field devices. In particular, the use of uncomplicated Simplex plug connectors for plastic fiber optics has had a considerable share in this development.

Bus line

Fiber optic cables with glass or plastic fibers are used as the transmission medium. Depending on the type of line used, glass fibers can handle connection distances of up to 15 km and plastic fibers up to 80 m.

Bus connection

Various connection techniques are available to connect bus stations to fiber optic conductors.

▷ OLM technology (*Optical Link Module*)

Similar to RS 485 repeaters, OLMs have two electrical channels which are functionally isolated and, depending on the model, one or two optical channels. OLMs are connected by an RS 485 line with the individual bus stations or the bus segment (see figure 1.7).

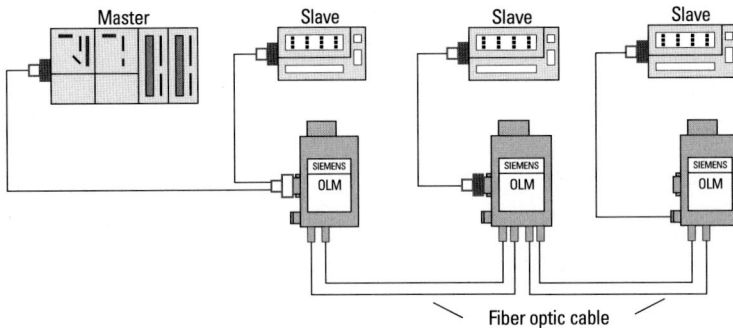

⊑ RS 485 bus connector
without terminating resistor

⊑ RS 485 bus connector
with terminating resistor

Figure 1.7
Example of a bus
configuration with
OLM technology

▷ OLP technology (*Optical Link Plug*)

OLPs can be used to connect very simple passive bus stations (slaves) with an optical single-fiber ring. OLPs are plugged directly onto the 9-pin sub D plug connector of the bus station. OLPs are powered by the bus stations and do not require their own power supplies. Note, however, that the $+5$ V part of the bus station's RS 485 interface must be able to provide a current of at least 80 mA (see figure 1.8).

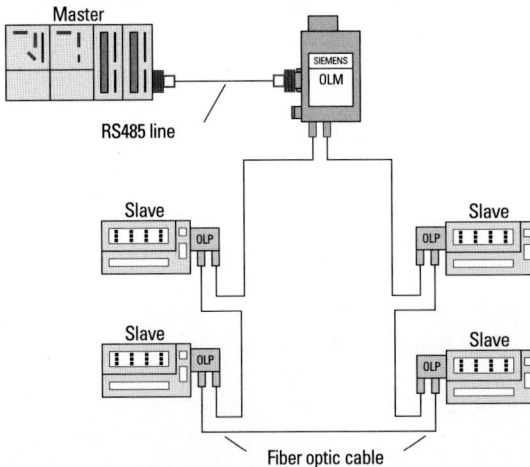

⊑ RS 485 bus connector with terminating resistor

Figure 1.8
Optical single-fiber ring with
OLP technology

Connecting an active bus station (master) to an OLP ring always requires an Optical Link Module.

▷ Integrated fiber optic cable connection

Direct connection of PROFIBUS nodes to the fiber optic cable using the fiber optic interface integrated in the device.

1.3.3 Physical Layer (Layer 1) for PA

PROFIBUS PA uses transmission technology in accordance with the IEC 1158-2 standard. This technology ensures intrinsic safety and bus powering of the field devices directly over the bus line. A bit-synchronous, Manchester-coded line protocol with dc-free transmission is used for data transmission (also known as H1 code). With Manchester-coded data transmission, a binary "0" is transmitted for the signal edge change from 0 to 1, and a binary "1" is transmitted for the signal edge change from 1 to 0. Data is transmitted by modulating $+/-$ 9 mA onto the base current I_B of the bus system (see figure 1.9). The transmission speed is 31.25 kbit/s. A twisted-pair, shielded or unshielded line is used as the transmission medium. The bus line is terminated at the segment ends by means of an RC passive line termination (see figure 1.10). Up to 32 bus stations can be connected on one PA bus segment. The maximum segment length depends to a great extent on the power supply, the type of line and the current consumption of the connected bus stations.

Bus line

A 2-core cable is required as the transmission medium for PROFIBUS PA. Its properties are not specified or standardized. However, the characteristics of the bus cable type determine the maximum expansion of the bus, the number of bus stations which can be connected and the sensitivity to electromagnetic interference.

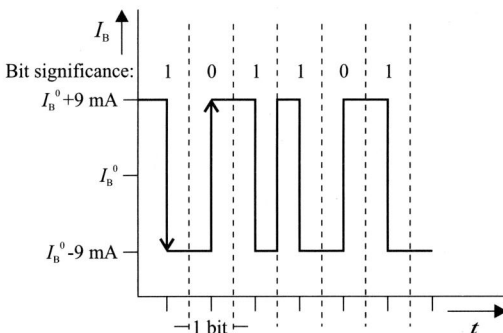

Figure 1.9
Data transmission with PROFIBUS PA by means of current modulation (Manchester-II code)

21

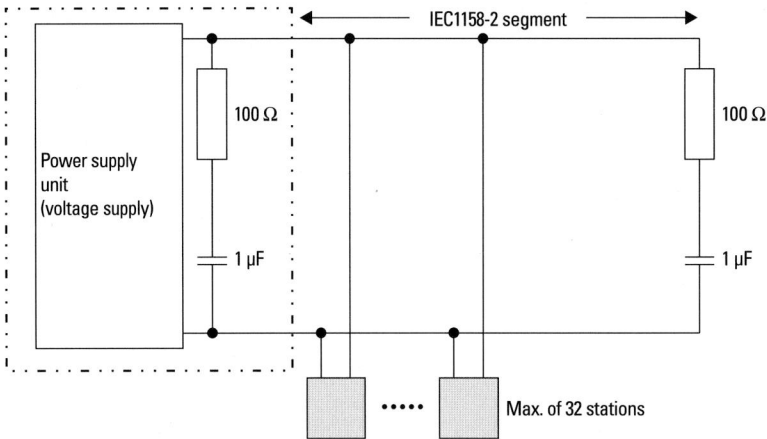

Figure 1.10   Layout of a PA bus segment

Therefore, the electrical and physical characteristics of several standard cable types have been defined in the DIN 61158-2 standard. This standard recommends four standard cable types for use with PROFIBUS PA, called type A to D (see table 1.4).

Table 1.4   Recommended cable types for PROFIBUS PA

	Type A (Reference)	Type B	Type C	Type D
Cable design	Twisted pair, shielded	One or more twisted pairs, total shield	Several twisted pairs, unshielded	Several non-twisted pairs, unshielded
Core cross section (rated)	0.8 mm^2 (AWG 18)	0.32 mm^2 (AWG 22)	0.13 mm^2 (AWG 26)	1.25 mm^2 (AWG 16)
Loop resistance (direct current)	44 Ω/km	112 Ω/km	264 Ω/km	40 Ω/km
Surge impedance at 31.25 kHz	100 Ω ± 20 %	100 Ω ± 30 %	**	**
Wave attentuation at 39 kHz	3 dB/km	5 dB/km	8 dB/km	8 dB/km
Capacitive asymmetry	2 nF/km	2 nF/km	**	**
Group distortion (7.9 to 39 kHz)	1.7 μs/km	**	**	**
Degree of covering of the shield	90 %	**	–	–
Recommended network size (incl. stub lines)	1900 m	1200 m	400 m	200 m

** Not specified

1.3.4 Fieldbus Data Link (Layer 2)

According to the OSI reference model, Layer 2 defines: bus access control (section 1.2), data security and processing of transmission protocols and telegrams. With PROFIBUS, Layer 2 is called the FDL Layer (*Fieldbus Data Link*).

Layer 2 telegram formats (figure 1.11) provide a high degree of transmission security. The call telegrams have a *Hamming Distance* of HD=4. HD=4 means that up to three simultaneously distorted bits can be detected in the data telegram. This is achieved by applying the regulations of the international IEC 870-5-1 standard, by selecting special start and end identifiers for the telegrams, by using gap-free synchronization, and by using a parity bit and a control byte. The following types of errors can be detected.

▷ Character format error (parity, overrun, framing error)

▷ Protocol error

▷ Start and end delimiter error

▷ Frame check byte error

▷ Telegram length error

Telegrams which were found to be faulty are automatically repeated at least once. The repetition of telegrams in Layer 2 can be set to a maximum of 8 ("retry" bus parameter). In addition to logical point-to-point data transmission, Layer 2 also allows multiple-point transmission with Broadcast and Multicast communication.

With Broadcast communication, an active station sends a message to all other stations (master and slaves). Receipt of the data is not acknowledged.

With Multicast communication, an active station sends a message to a group of stations (master and slaves). Receipt of the data is not acknowledged.

The data services offered by Layer 2 are listed in table 1.5.

Table 1.5 PROFIBUS transmission services

Service	Function	DP	PA	FMS
SDA	Send Data with Acknowledge			×
SRD	Send and Request Data with Acknowledge	×	×	×
SDN	Send Data without Acknowledge	×	×	×
CSRD	Cyclic Send and Request Data with Acknowledge			×

Format with fixed information field length

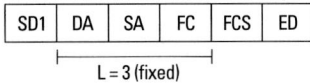

SD1	DA	SA	FC	FCS	ED

L = 3 (fixed)

Format with fixed information field length with data

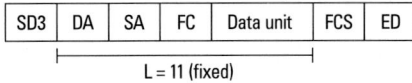

SD3	DA	SA	FC	Data unit	FCS	ED

L = 11 (fixed)

Format with variable information field length

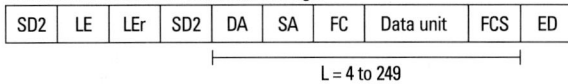

SD2	LE	LEr	SD2	DA	SA	FC	Data unit	FCS	ED

L = 4 to 249

Short acknowledgment

SC

Token telegram

SD4	DA	SA

L	Length of the information field
SC (Single Character)	Single character; only used for acknowledgment
SD1 (Start Delimiter) to SD4	Start byte; distinguishes between different telegram formats
LE/LEr (LEngth)	Length byte, indicates the length of the iniformation field in telegrams of variable length
DA (Destination Address)	Destination address byte; indicates which station is to receive the message
SA (Source Address)	Source address byte, indicates which station is to transmit the message
FC (Frame Control)	Control byte; contains details about the service used for this message and the priority of the message
Data Unit	Data unit; contains the useful information of the telegram and details about an extended address, if applicable
FCS (Frame Check Sequence)	Check byte; contains a telegram check sum which is obtained by ANDing all telegram elements without using a carry bit
ED (End Delimiter)	End byte; indicates the end of the telegram

Figure 1.11   PROFIBUS telegram formats

PROFIBUS DP and PROFIBUS PA each use a specific subset of Layer 2 services. PRO-FIBUS DP, for example, exclusively uses the SRD and SDN services.

Higher order layers call these services through the SAPs (*Service Access Point*) of Layer 2. With PROFIBUS FMS, these service access points are used to address the logical communication relationships. With PROFIBUS DP and PROFIBUS PA, each service access point used is assigned to a defined function. All active and passive stations allow the simultaneous use of several service access points. We distinguish between SSAPs (*Source Service Access Point*) and DSAPs (*Destination Service Access Point*).

1.3.5 Application Layer (Layer 7)

Layer 7, the Application Layer of the ISO/OSI reference model, provides the communication services required by the user. The Application Layer of PROFIBUS consists of the FMS interface (*Fieldbus Message Specification*) and the LLI interface (*Lower Layer Interface*).

FMS profiles

FMS profiles were defined by the PNO (PROFIBUS International) to adjust the FMS communication services to the functional range actually required and to define the device functions with respect to the actual application. These FMS profiles ensure that devices of different manufacturers have the same communication functionality. The following FMS profiles have been defined to date.

Communication between programmable controllers (3.002)

This communication profile specifies which FMS services will be used for communication between programmable controllers (PLCs). Based on defined controller classes, this profile specifies which PLC must be able to support which services, parameters and data types.

Profile for building services automation (3.011)

This profile is a sector-related profile and serves as the basis for many public calls for bids in the sector of building services automation. It describes how monitoring, open and closed-loop control, operator control, alarms and archiving of building services automation systems are to be handled by FMS.

Low voltage switch gear devices (3.032)

This profile is a sector-oriented FMS application profile. It specifies the response of low-voltage switch gear devices during data communication with FMS.

DP user interface and DP profiles

PROFIBUS DP only uses Layers 1 and 2. The User Interface defines the available application functions, and the system and device behavior of the various types of PROFIBUS DP devices.

The only task of the PROFIBUS DP protocol is to define how user data is transmitted from one station to another over the bus. There is no evaluation of the transmitted user data by the transmission protocol. This is the task of the DP profiles. Precisely specified application-related parameters and the use of profiles make it easy to mix individual DP components of different manufacturers. The following PROFIBUS DP profiles have been specified to date:

Profile for NC/RC (3.052)

The profile describes how handling and installation robots are controlled by PROFIBUS DP. Precise sequence diagrams describe the motion and program control of the robots from the point of view of the higher-level automation plant.

Profile for encoder (3.062)

The profile describes how shaft, shaft-angle and linear encoders with single-turn or multi-turn resolution are coupled to PROFIBUS DP. Two classes of devices define the basic functions and advanced functions, such as scaling, alarm handling and extended diagnostics.

Profile for variable-speed drives (3.072)

Leading manufacturers of drive technology joined forces to develop the PROFIDRIVE profile. The profile specifies how to define drive parameters and how to transmit setpoints and actual values. This makes it possible to use and mix drives of different manufacturers.

The profile contains the specifications required for the operating modes "speed control" and "positioning." It specifies the basic drive functions while leaving sufficient freedom for application-related expansions and further developments. The profile contains an image of the DP application functions or, alternatively, of the FMS application functions.

Profile for operator control and process monitoring, HMI
(Human Machine Interface) (3.082)

The profile for simple HMI devices defines the connection of such devices to higher-level automation components through PROFIBUS DP. For data communication, this profile makes use of the expanded set of PROFIBUS DP functions.

Profile for error-proof data transmission with PROFIBUS DP (3.092)

This profile defines additional data security mechanisms for communication with fail-safe devices, such as Emergency OFF. The security mechanism specified by this profile has been approved by TÜV (German Technical Inspectorate) and BIA.

1.4 Bus Topology

1.4.1 RS 485

The PROFIBUS system consists of a linear bus structure which is actively terminated on both sides. This is also known as an RS 485 bus segment. Based on the RS 485 standard, up to 32 RS 485 stations (also referred to as "nodes") can be connected to one bus segment. Whether master or slave, each station connected to the bus represents an RS 485 current load.

RS 485 is the least expensive and also the most frequently used transmission technique with PROFIBUS.

Repeaters

A PROFIBUS system that is to accommodate more than 32 stations must be divided into several bus segments. These individual bus segments with up to 32 stations each are connected to each other by repeaters (also referred to as line amplifiers). The repeater amplifies the level of the transmission signal. The EN 50 170 standard does not provide for time regeneration of the bit phases within the transmission signal (signal regeneration) by the repeater. Due to the resulting distortion and delay of the bit signals, EN 50 170 limits the number of repeaters connected in series to three. These repeaters act as pure line amplifiers. However, in practice, signal regeneration has been implemented on the repeater circuits. Therefore, the number of repeaters which may be connected in series depends on the particular repeater and manufacturer. For example, up to 9 repeaters of type 6ES7 972-0AA00-0XA0 from Siemens may be connected in series.

The maximum distance between two bus stations depends on the baud rate. Table 1.6 specifies the values for a repeater of type 6ES7 972-0AA00-0XA0.

Table 1.6
Maximum expansion of a PROFIBUS configuration with 9 repeaters connected in series, as a function of the baud rate

Baud rate (kbit/s)	9.6 to 187.5	500	1,500	12,000
Total length of all segments in m	10,000	4,000	2,000	1,000

Segment 1 ... Segment 2

A1	A2
B1	B2
A1	A2
B1	B2

Logic

5V — 24V — 1 M — 5V — 24V — 1 M

PG/OP socket

L+ (24 V)	L+ (24 V)
M	M
A1	
B1	PE
5V	M 5.2
M 5V	

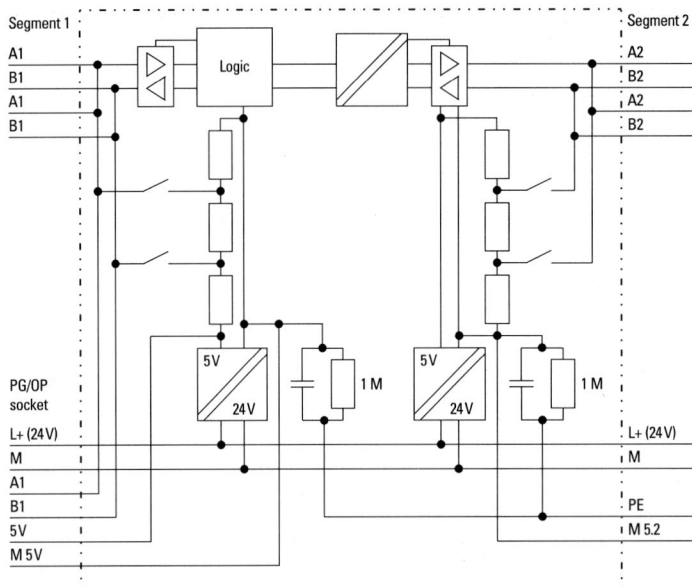

Figure 1.12   Block diagram of the RS 485 repeater type 6ES7 972-0AA00-0XA0

The block diagram shown in figure 1.12 describes the characteristics of the RS 485 repeater.

▷ Bus segment 1, PG/OP socket and bus segment 2 are galvanically isolated from each other.

▷ The signals between bus segment 1, PG/OP socket and bus segment 2 are amplified and regenerated.

▷ The repeater has connectable terminal resistors for bus segments 1 and 2.

▷ By removing jumper M/PE, the repeater can be operated ungrounded.

Only through the use of repeaters can the maximum possible number of stations be achieved in a PROFIBUS configuration. In addition, repeaters can be used to implement "tree" and "star" bus structures. An ungrounded layout is also possible. In this type of bus structure, the bus segments are isolated from each other. In this case you have to use a repeater and an ungrounded 24 V power supply (see figure 1.13).

For the RS 485 interface, a repeater is an additional load. Therefore, the maximum number of bus stations which can be operated in one bus segment must be reduced by one for each RS 485 repeater used. So, if the bus segment contains one repeater, you can operate up to 31 bus stations in this segment. However, the number of repeaters in the overall bus configuration has no effect on the maximum number of bus stations because the repeater does not occupy a logical bus address.

Stub lines

Direct connection of bus stations, for example on the 9-pin sub D plug connectors of the bus connection plugs, creates stub lines in the linear structure of the bus system.

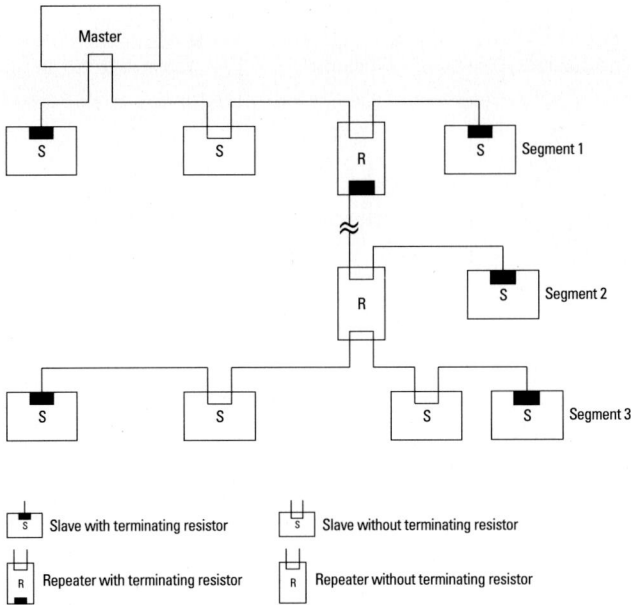

Figure 1.13   Bus configuration with repeaters

Although the EN 50 170 standard states that, at a transmission speed of 1500 kbit/s, stub lines shorter than 6.6 m per segment are permitted, it is usually best to avoid stub lines at the outset when the bus system is configured. An exception to this rule is the use of stub lines for temporarily connected programming units or diagnostic tools. Depending on their number and length, stub lines can cause line reflections which may interfere with telegram communication. Stub lines are not permitted for transmission speeds higher than 1500 kbit/s. In networks with stub lines, programming units and diagnostic tools may only be connected to the bus by the "active" bus connection lines.

1.4.2 Fiber Optics

Fiber optics for data transmission have paved the way for a new bus structure – the ring structure – in addition to the linear, tree or star bus structure known so far. Optical Link Modules (OLMs) can be used to implement both single-fiber rings and redundant optical double-fiber rings (see figure 1.14). In single-fiber rings, the OLMs are connected with each other by Simplex fiber optic cables. If a fault occurs, say, the fiber optic cable line is interrupted or an OLM fails, then the entire ring breaks down. In a redundant fiber optic ring, the OLMs are interconnected by means of two Duplex fiber optic cables each. They are therefore able to react if one of the two fiber optic lines breaks down, and automatically switch the bus system to a linear structure. Appropriate signaling contacts indicate the fault of the transmission line and pass this information on for further processing. As soon as the fault in the fiber optic line is eliminated, the bus system returns to the normal state of a redundant ring.

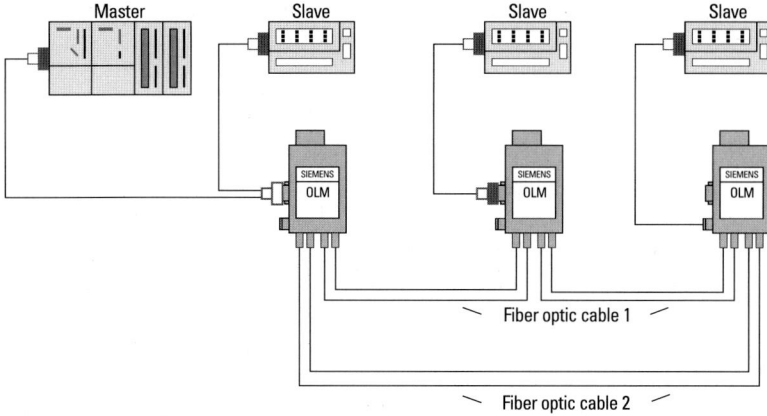

RS 485 bus connector with terminating resistor

RS 485 bus connector without terminating resistor

Figure 1.14   Redundant double-fiber ring

1.4.3 Topology acc. to IEC 1158-2 (PROFIBUS PA)

Using the PROFIBUS PA protocol, you can implement a linear, tree and star bus structure, or a combination of these. The number of bus stations which can be operated on one
bus segment depends on the power supply used, the current consumption of the bus stations, the bus cable used and the size of the bus system. Up to 32 stations can be connected to one bus segment. To increase system availability, bus segments can be backed
up by a redundant bus segment. A segment coupler (figure 1.15) or a DP/PA link is used
to connect a PA bus segment to a PROFIBUS DP bus segment.

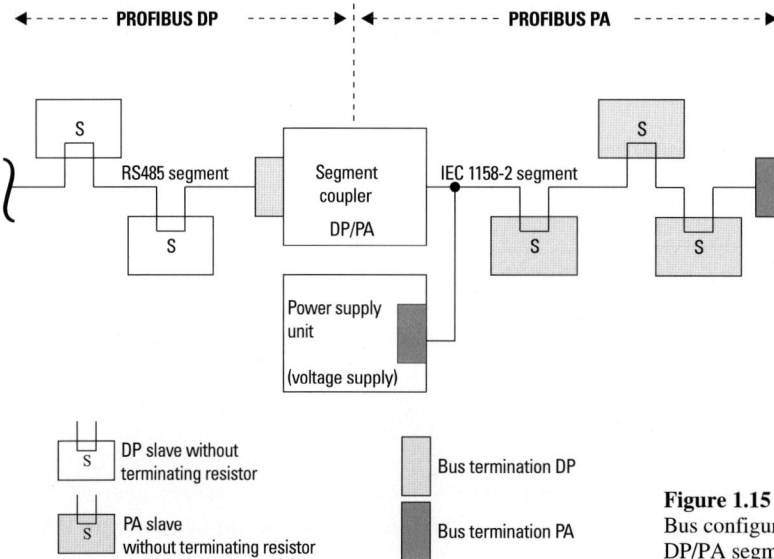

DP slave without terminating resistor

PA slave without terminating resistor

Bus termination DP

Bus termination PA

Figure 1.15
Bus configuration with
DP/PA segment coupler

1.5 Bus Access Control in a PROFIBUS Network

The bus access control of PROFIBUS meets two requirements that are vital for automated industrial and manufacturing processes, which after all are the principal application areas of fieldbus technology. On the one hand, communication between equal programmable controllers or PCs requires that each bus station (node) receives sufficient
opportunity to process its communication task within the defined period. Data traffic
between a complex PLC or PC and simple distributed process I/O peripherals, on the
other hand, must be fast and cause as little protocol overhead as possible.

PROFIBUS achieves this by using a hybrid bus access control mechanism. It consists of
a decentralized *token passing* procedure for communication between the *active* nodes
(master), and a centralized *master-slave* procedure for communication between the *active*
and the *passive* nodes.

When an active node (bus station) has the token, it takes over the master function on the
bus to communicate with both passive and active nodes. The exchange of messages on
the bus is organized by means of node addressing. Each PROFIBUS node is given an
address which must be unique throughout the entire bus system. The maximum usable
address range within a bus system lies between addresses 0 and 126. This means that the
bus system can have a maximum of 127 nodes (bus stations).

This method of bus access control allows the following system configurations:

▷ Pure master-master system (token passing)

▷ Pure master-slave system (master-slave)

▷ Combination of the two procedures

The bus access procedure of PROFIBUS is not dependent on the transmission medium
used. Whether the network uses copper cables or fiber optics is irrelevant. The
PROFIBUS bus access control complies with the token bus procedure and the master-
slave procedure specified in the European EN 50 170 standard, Volume 2.

1.5.1 Token Bus Procedure

The active nodes connected to the PROFIBUS network form a logical *token ring* in ascending order of their bus addresses (see figure 1.16). A token ring is a succession of active nodes in which a control token is always passed from one station to the next. The token provides the right to access the transmission medium, and is passed between the active nodes with a special token telegram. The active node with bus address HSA (*Highest Station Address*) is an exception. This node only passes the token to the active node with the lowest bus address to close the logical token ring again.

The time required for one rotation of the token to all active nodes is called the token rotation time. The adjustable token time Ttr (*Time Target Rotation*) is used to specify the maximum time permitted to the fieldbus system for one token rotation.

In the bus initialization and startup phase, the bus access control (also known as MAC-Medium Access Control) establishes the token ring by recognizing the nodes that are active. To manage the control token, the MAC procedure first automatically determines the addresses of all active nodes on the bus, and records them together with its own node address in the LAS (*List of Active Stations*). Particularly important for token management are the addresses of the PS node (*Previous Station*) from which the token is received, and the NS node (*Next Station*) to which the token is passed. The LAS is also required during running operation to remove a faulty *active* node from the ring, or add a new node to the ring, without disturbing data communication on the bus.

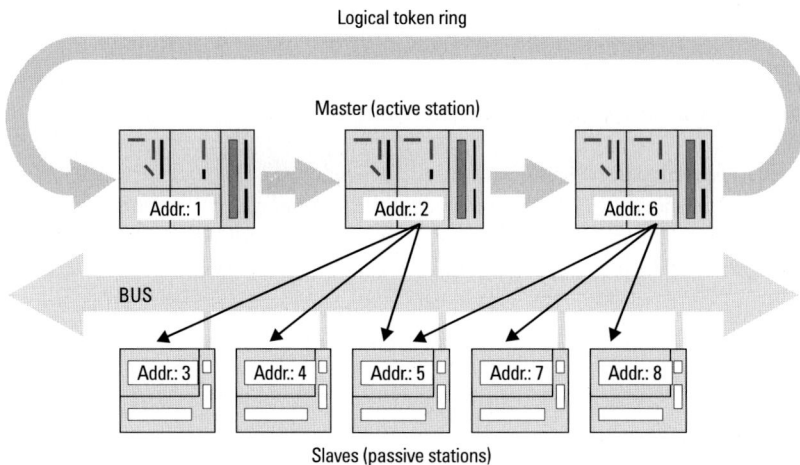

Figure 1.16   Token bus procedure

1.5.2 Master-Slave Procedure

A network that has several passive nodes, but whose logical token ring consists of only one active node, is a pure master-slave system (see figure 1.17).

The master-slave procedure permits the master – the active node – that currently has the right to send, to address the slave devices which are assigned to it. These slaves are the passive nodes. The master can send messages to the slaves or fetch messages from the slaves.

The typical standard PROFIBUS DP bus configuration is based on this bus access procedure. An active node (master) cyclically exchanges data with the passive nodes (DP slaves).

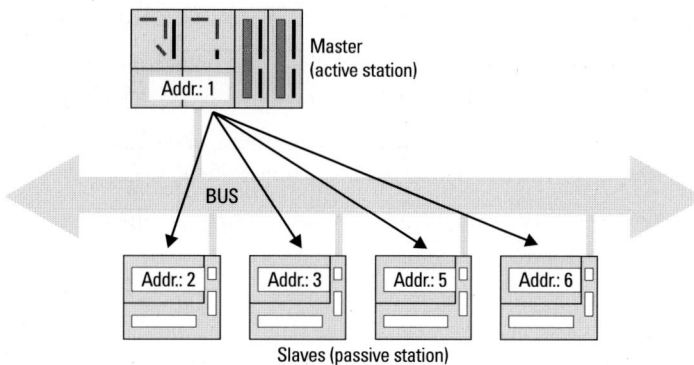

Figure 1.17   Master-slave procedure

1.6 Bus Parameters

Faultless functioning of a PROFIBUS network is only guaranteed if the bus parameters set have been coordinated with each other. The bus parameters set at a node must also be set at every other node in the same network so that they are identical in the overall PROFIBUS network. In general, the bus parameters depend on the selected data transfer rate and are specified by the configuring tool in each case. These parameter sets may only be modified by experienced personnel. Here are the most important bus parameters and their definitions:

Ttr: The *target rotation* time is the maximum time provided for a token to pass round all bus nodes. During this time, all active nodes receive the authorization (token) once to transmit data on PROFIBUS. The difference between the target rotation time and the actual time the token spends at a node determines how much time is available to the other active nodes for sending message frames.

GAP factor: The GAP factor defines the number of token rotations after which an attempt is made to include a new, active node in the logical token ring.

RETRY limit: The RETRY limit defines how often a message frame is repeated following an incorrect acknowledgement or timeout.

Min_TSDR: The *minimum station delay responder* is the minimum time a passive node must wait to be allowed to respond to a message frame.

Max_TSDR: The *maximum station delay responder* is the maximum time a passive node is allowed to respond to a message frame.

Tslot: The *slot time* defines the maximum time a sender is allowed to wait for a response from the addressed node.

Tset: The *setup time* is the period that may elapse between receiving a message frame and the node's response to it.

Tqui: The *quiet time for modulator* describes how long a transmitting node is allowed for switching to receive after sending a message frame.

Tid1: *Idle time 1* defines the earliest possible time after which a transmitting node is allowed to send a message frame again after receiving a response.

Tid2: *Idle time 2* is the time a node must wait after sending an unacknowledged message frame (broadcast) before sending another message frame.

Trdy: The *ready-time* describes the time after which a transmitting node can receive a response frame.

All bus parameters thus describe times that must be coordinated precisely with each other. The unit for specifying these bus parameters is the tBIT (*time_bit*). A tBIT is the bus rotation time for one bit and is also referred to as the bit rotation time. This time depends on the data transfer rate and is calculated as follows:

$tBIT$ = 1/data transfer rate (bit/s)

For example, the bit rotation time for a data transfer rate of 12 Mbit/s is 83 ns, and the bit rotation time for a data transfer rate of 1.5 Mbit/s is 667 ns.

2 Types of Bus Devices and
Data Communication with PROFIBUS DP

Introduction

The PROFIBUS DP protocol is designed for the high-speed data communication required for distributed I/O and field devices in automated industrial plants. The typical DP configuration has a mono-master structure (see figure 2.1). Communication between the DP master and the DP slaves is based on the master-slave principle. This means that the DP slaves may only become active on the bus when requested by the master. The DP slaves are addressed in succession by the DP master using a polling list. User data between the DP master and the DP slaves is continuously (i.e., cyclically) exchanged irrespective of the contents of the user data. Figure 2.2 shows how the polling list is processed on the DP master. A message cycle between the DP master and a DP slave consists of a request frame (polling telegram) issued by the DP master, and the related acknowledgment or response frame returned by the DP slave.

Due to the characteristics in Layer 1 and Layer 2 of PROFIBUS nodes as specified by the EN 50170 standard, a DP system may also have a multi-master structure. In practice, this may mean that several DP master nodes are connected to one bus line. The coexistence of DP master/slaves, FMS master/slaves and additional active or passive nodes on one bus line is also possible (see figure 2.3).

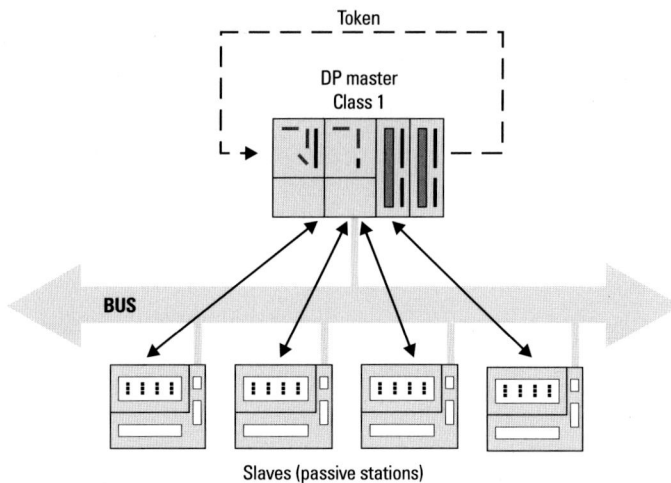

Figure 2.1   DP mono-master structure

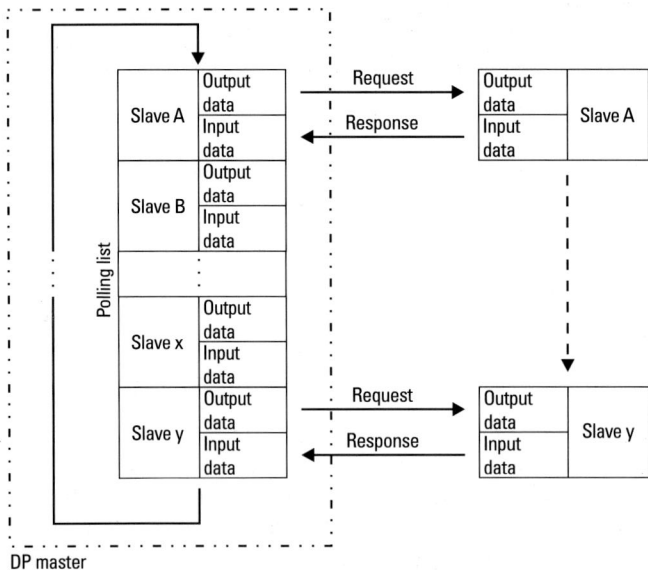

Figure 2.2   Processing the polling list on the DP master

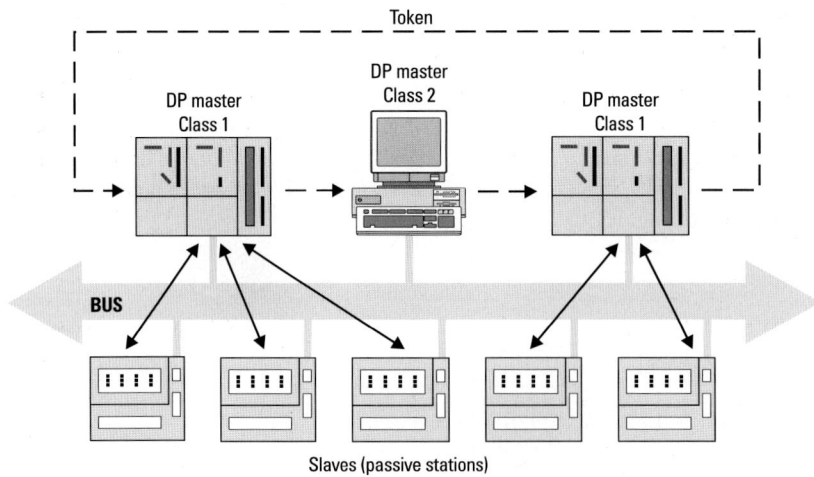

Figure 2.3   PROFIBUS multi-master structure

2.1 Types of Bus Devices

2.1.1 DP Master (Class 1)

This DP master cyclically exchanges user data with the DP slave. The Class 1 DP Master executes tasks using the following protocol functions.

• *Set_Prm* and *Chk_Cfg*

The DP master uses these functions in the phases of startup, restart and data transfer to transmit the parameter sets to the DP slaves. It transmits all parameters, irrespective of whether they apply globally to the entire bus or are of some specific importance. The number of input and output data bytes for the particular DP slave is defined during configuration.

• *Data_Exchange*

This function handles the cyclic exchange of input and output data with the assigned DP slave.

• *Slave_Diag*

This function reads the diagnostic information of the DP slave during startup or during the cyclic exchange of user data.

• *Global_Control*

The DP master uses control commands to inform the DP slaves of its operational status. In addition, control commands can be sent to individual slaves or to specified groups of DP slaves to synchronize the output and input data (Sync and Freeze command).

2.1.2 DP Slave

A DP slave only exchanges user data with the DP master that was responsible for loading its parameters and configuring it. A DP slave is able to report local diagnostic interrupts and process interrupts to the DP master.

2.1.3 DP Master (Class 2)

Class 2 DP masters are devices such as programming units, and diagnostic and bus management devices. In addition to the class 1 functions already described, class 2 DP masters usually support the following special functions:

• *RD_Inp* and *RD_Outp*

These functions read the input and output data of DP slaves at the same time as data communication with the class 1 DP master takes place.

• *Get_Cfg*

This function reads the current configuration data of a DP slave.

• *Set_Slave_Add*

This function permits the DP master to assign a new bus address to a DP slave, provided that the slave supports this method of address definition.

In addition, the class 2 DP master provides a number of functions for communication with the class 1 DP master.

2.1.4 DP Combination Devices

It is possible to combine several of the DP device types "Class 1 DP master," "Class 2 DP master" and "DP slave" in one and the same hardware module. In practice, you will find this quite frequently. The following are some typical device combinations:

- Class 1 DP master combined with Class 2 DP master

- DP slave with Class 1 DP master

2.2 Data Communication between the Various Types of DP Devices

2.2.1 DP Communication Relationships and DP Data Exchange

With the PROFIBUS DP protocol, the initiator of a communication job is called a requester and the appropriate communication partner is called the responder. All request telegrams of the Class 1 DP master are processed in Layer 2 by the "high priority" telegram service class. With one exception, the response telegrams sent by the DP slaves use the "low-priority" telegram service class in Layer 2. The DP slave can inform the DP master that current diagnostic interrupts or status events are pending. It does this by changing just once the *Data_Exchange* response telegram service class from "low priority" to "high-priority." Transmission of the data is connectionless through *one-to-one* or *one-to-many* connections (control commands and cross communication only). Table 2.1 lists the communication capabilities of the DP masters and the DP slave, arranged according to requester and responder function.

Table 2.1 Communication relationships between the various types of DP device

Function/Service (acc. to EN 50170)	DP slave		DP master (Class 1)		DP master (Class 2)		Through SAP Number	Through Layer 2 Service
	Requ	Resp	Requ	Resp	Requ	Resp		
Data_Exchange		M	M		O		Default-SAP	SRD
RD_Inp		M			O		56	SRD
RD_Outp		M			O		57	SRD
Slave_Diag		M	M		O		60	SRD
Set_Prm		M	M		O		61	SRD
Chk_Cfg		M	M		O		62	SRD
Get_Cfg		M			O		59	SRD
Global_Control		M	M		O		58	SDN
Set_Slave_Add		O			O		55	SRD
M-M-Communication			O	O	O	O	54	SRD/SDN
DP V1 Services		O	O		O		51/50	SRD

Requ = Requester, Resp = Responder, M = Mandatory function, O = Optional function

2.2.2 Initialization Phase, Restart and User Data Communication

As shown in figure 2.4, the DP master has to define the parameters of the DP slave and configure it before it can exchange user data with the slave device. It does this by first checking whether the DP slave reports on the bus. If so, the DP master checks readiness of the DP slave by requesting the slave's diagnostic data. When the DP slave reports that it is ready for parameter definition, the DP master loads the parameter set and configuration data. The DP master again requests diagnostic data from the slave to find out whether it is ready. Only then does the DP master start to cyclically exchange user data with the DP slave.

Parameter data (Set_Prm)

The parameter set contains important local and global parameters, properties and functions intended for the DP slave. You will usually use the configuration tool provided with the DP master for specifying and configuring the slave parameters. With the direct configuration method, you fill in dialog boxes offered by the graphical user interface of the configuration software. Indirect configuration consists in accessing existing parameters and DP slave-related GSD data (*Geräte Stamm Daten*, device master data) using the configuration tool. The layout of the parameter telegram consists of a part specified by the

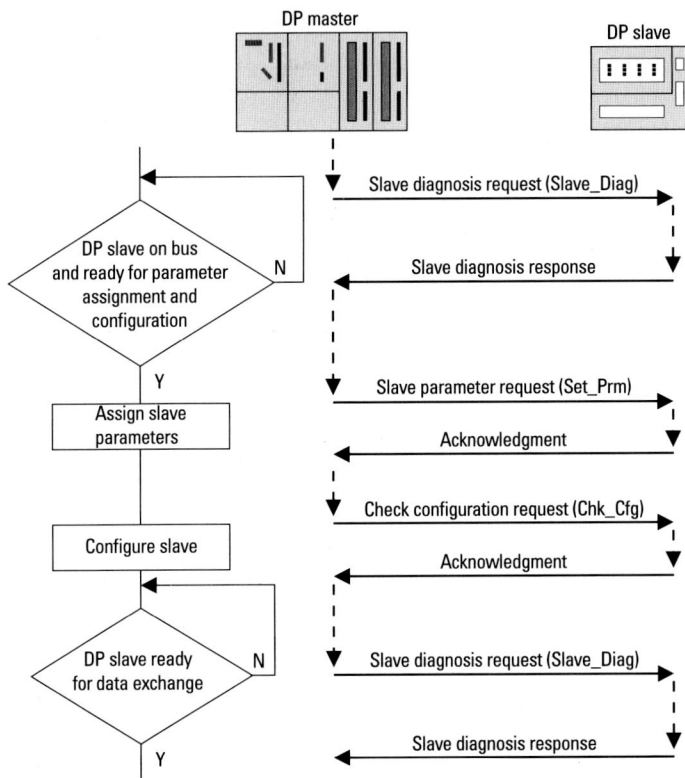

Figure 2.4   Principal sequence of the initialization phase of a DP slave

39

EN 50 170 standard and, if required, a DP slave and manufacturer-specific part. The length of the parameter telegram may not exceed 244 bytes. The most important contents of the parameter telegram are listed below.

• *Station Status*

Station Status contains slave-related functions and settings. For example, it specifies whether or not watchdog monitoring is to be activated. It also defines whether access to the DP slave by other DP masters is to be enabled or disabled, and, if provided for in the configuration, whether Sync or Freeze control commands are to be used with this slave.

• *Watchdog*

The watchdog detects the failure of a DP master. If the watchdog is enabled and the DP slave detects the failure of the DP master, the local output data is deleted or secured in a defined state (substitute values are transferred to the outputs). A DP slave can be operated on the bus with or without the watchdog. Based on the bus configuration and the selected transmission speed, the configuration tool suggests a watchdog time which can be used for the configuration. Please also see "bus parameters."

• *Ident-Number*

The Ident-Number of the DP slave is assigned during certification by the PNO ("*Profi-bus-Nutzer-Organisation*" = PROFIBUS International). The Ident-Number of the DP slave is stored in the device master file. A DP slave will only accept a parameter telegram if the Ident-Number received with this telegram corresponds to its own. This prevents accidental incorrect parameter definitions on the slave device.

• *Group-Ident*

Group-Ident permits DP slaves to be combined in groups for the Sync and Freeze control commands. Up to 8 groups are permitted.

• *User-Prm-Data*

DP slave parameter data (User-Prm-Data) specifies application-related data for the DP slave. For example this can include default settings or controller parameters.

Configuration data (Chk_Cfg)

In the configuration data telegram, the DP master transmits identifier formats to the DP slave. These identifier formats inform the DP slave about scope and structure of the input/output area to be exchanged. These areas (also referred to as "modules") are defined in the form of byte or word structures (identifier formats) agreed by the DP master and the DP slave. The identifier format allows you to specify the input or output areas, or the input and output areas for each module. These data areas can have a maximum size of 16 bytes/words. When you define the configuration telegram, you have to consider the following characteristics, based on the type of the DP slave device:

- The DP slave has a fixed input and output area (e.g., block I/O ET200B).

- Depending on the configuration, the DP slave has a dynamic input/output area (e.g., modular I/O such as for ET200M or drives).

- The input/output area of the DP slave is specified by means of special identifier formats which depend on the DP slave and manufacturer (e.g. S7 DP slaves such as ET200B-Analog, DP/AS I-Link and ET200M).

Input and output data areas which contain coherent information but cannot be placed in a byte or word structure, are considered as "consistent" data. This includes parameter areas for closed-loop controllers or parameter sets for drive control, for example. Using special identifier formats (DP slave and manufacturer-related), you can specify the input and output areas (modules) with a length of up to 64 bytes/words.

The input and output data areas (modules) which can be used by the DP slave are stored in the device master file (GSD file). They will be suggested to you by the configuration tool when you configure the DP slave.

Diagnostic data (Slave_Diag)

By requesting the diagnostic data, the DP master checks during the startup phase whether the DP slave exists and is ready to receive the parameter information. The diagnostic data supplied by the DP slave consists of a diagnostic part according to the EN 50 170 standard and, if present, specific DP slave diagnostic information. The DP slave transmits diagnostic data to inform the DP master about its operational state and, in the event of an error, the cause of the error. A DP slave is able to generate a local diagnostic interrupt to Layer 2 of the DP master, using the "high-prio" telegram of the Data_Exchange Response telegram in Layer 2. In response, the DP master requests the diagnostic data for evaluation. If there are no current diagnostic interrupts, then the Data_Exchange Response telegram has a "low-priority" identifier. However, the diagnostic data of a DP slave can always be requested by a DP master even when no special report of diagnostic interrupts have been reported.

User data (Data_Exchange)

The DP slave checks the parameter and configuration information received from the DP master. If there are no errors and the settings requested by the DP master are permitted, the DP slave transmits diagnostic data to report that it is ready for the cyclic exchange of user data. Starting now, the DP master exchanges the configured user data with the DP slave (see figure 2.5). During the exchange of user data, the DP slave only reacts to the *Data_Exchange* request telegram sent by the Class 1 DP master that was responsible for its parameter definition and configuration. Other user data telegrams is rejected by the DP slave. The user data contains no additional control or structure characters to describe the transmitted data. This means that only useful data is transmitted.

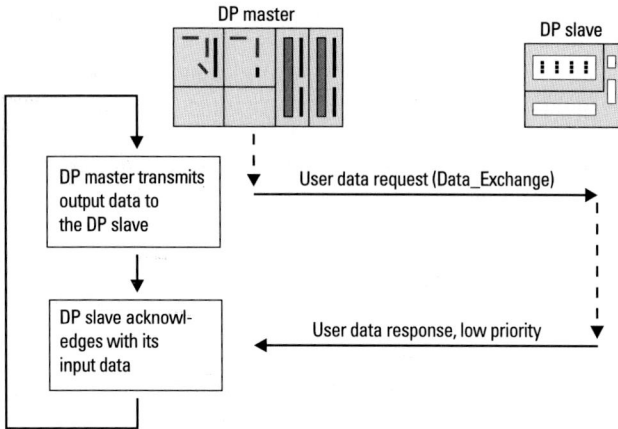

Figure 2.5   DP slave during cyclic exchange of user data with the DP master

As shown in figure 2.6, the DP slave can tell the DP master that current diagnostic inter-rupts or status messages exist by changing the telegram service class in the response from "low-priority" to "high-priority." The DP master then makes one request of the actual diagnostic or status information which is sent by the DP slave in a diagnostic telegram. After the diagnostic data has been fetched, DP slave and DP master return to exchanging user data. Using request/response telegrams, the DP master and the DP slave can ex-change up to 244 bytes of user data in both directions.

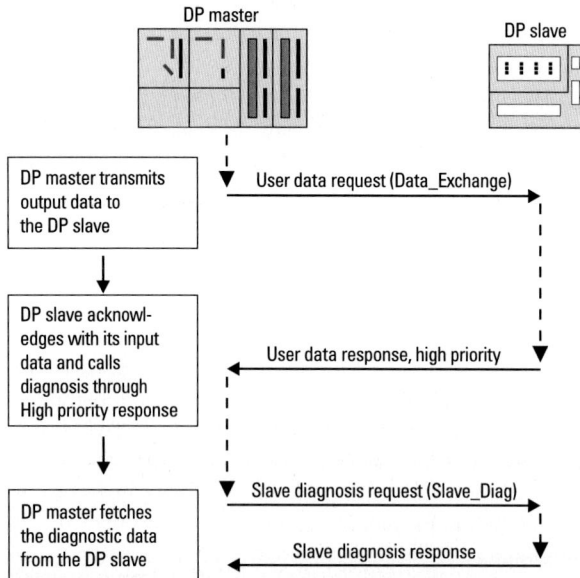

Figure 2.6   DP slave reports a current diagnostic interrupt

2.3 PROFIBUS DP Cycle

2.3.1 Setup of a PROFIBUS DP Cycle

Figure 2.7 shows the setup of a DP cycle in a DP mono-master bus system. The DP cycle comprises a fixed part and a variable part. The fixed part is made up of the cyclic telegrams containing the bus access control (token management and station status) and the I/O data communication (Data_Exchange) with the DP slaves.

The variable part of the DP cycle is made up of a number of event-controlled, non-cyclic telegrams. This non-cyclic part of the telegram includes the following:

• Data communication during the initialization phase of a DP slave

• DP slave diagnostic functions

• Class-2 DP master communication

• DP master, master communication

• Layer-2 controlled telegram repetitions during malfunctions (Retry)

• Non-cyclic data communication acc. to DPV1

• PG online functions

• HMI functions

Depending on how many non-cyclic telegrams occur in the current DP cycle, the DP cycle increases accordingly.

Thus, a DP cycle always consists of a fixed cyclic time and, if present, an event-controlled, variable, non-cyclic number of telegrams.

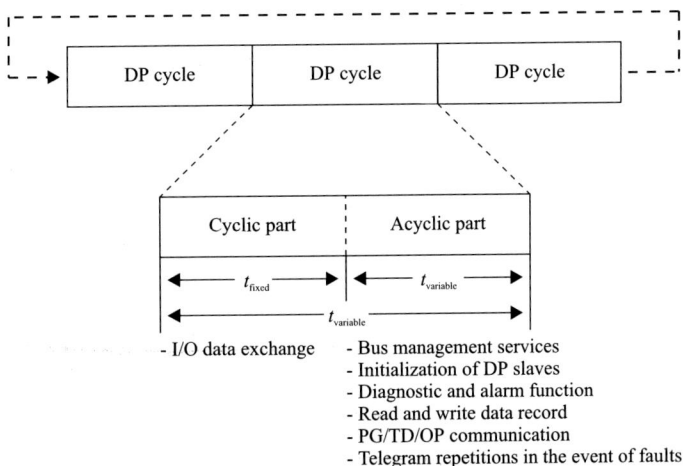

Figure 2.7   Principal setup of a PROFIBUS DP cycle

2.3.2 Setup of a Constant PROFIBUS DP Cycle

For certain applications in the automation sector, a constant DP bus cycle time and thus a constant I/O data exchange is advantageous. This applies particularly to the field of drive control. Synchronization of several drives would require a constant bus cycle time, for example. Note that a constant bus cycle is also often referred to as an "equidistant" bus cycle.

In contrast to the normal DP cycle, a certain time is reserved for non-cyclic communication during a constant DP cycle of the (constant bus cycle) DP master. As shown in figure 2.8, the DP master ensures that this reserved time is not exceeded. It only permits a certain number of non-cyclic telegram events. If the reserved time is not needed, the DP master bridges the still missing difference to the selected constant bus cycle time by sending telegrams to itself, thus creating a pause. This ensures that the reserved constant bus cycle time is adhered to down to the microsecond.

The time for the constant DP bus cycle is specified in the STEP 7 configuration software. The (default) time suggested by STEP 7 depends on the configured system and takes into account a certain typical portion of non-cyclic services. You can of course change the constant bus cycle time suggested by STEP 7.

To date, a constant DP cycle time can only be set in mono-master mode.

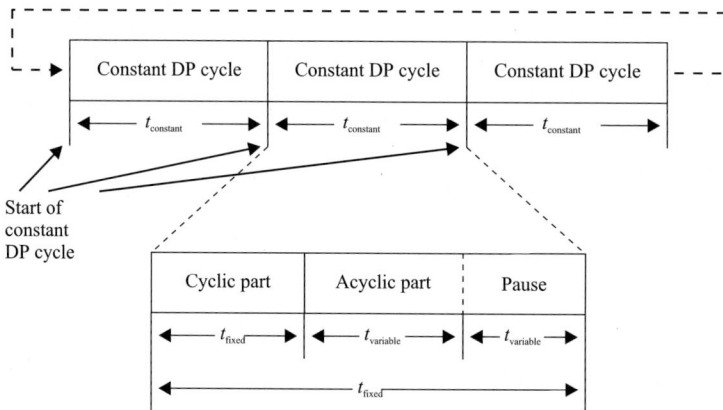

Figure 2.8   Setup of a constant PROFIBUS DP cycle

2.4 Data Exchange by means of Cross Communication

Cross communication (also referred to as "direct communication") is another method of data communication with PROFIBUS DP in SIMATIC S7 applications. During cross communication, the DP slave does not respond to the DP master with a *one-to-one* telegram (slave -> master), but with a special *one-to-many* telegram (slave -> nnn). This means that the input data of the slave contained in the response telegram is available not only to the related master, but also to all DP nodes on the bus which support this function.

With cross communication, the communication relationships "master-slave" and "slave-slave" are possible, but they are not supported by all models of SIMATIC S7 DP master and slave devices. Use the STEP 7 software to define the type of relationship. A combination of both methods of communication is allowed.

2.4.1 Master-Slave Relationship with Cross Communication

Figure 2.9 shows the master-slave relationships you can set up in a DP multi-master system consisting of three DP masters and four DP slaves. All slaves shown in this figure send their input data as a *one-to-many* telegram. DP master A, to which slaves 5 and 6 are assigned, also uses this telegram to receive the input data of slaves 7 and 8. Similarly, DP master B, to which slaves 7 and 8 are assigned, also receives the input data of slaves 5 and 6. Although the DP master shown as master C has no slaves assigned to it, it receives the input data of all slaves operated on the bus system (i.e., the data of slaves 5, 6, 7 and 8).

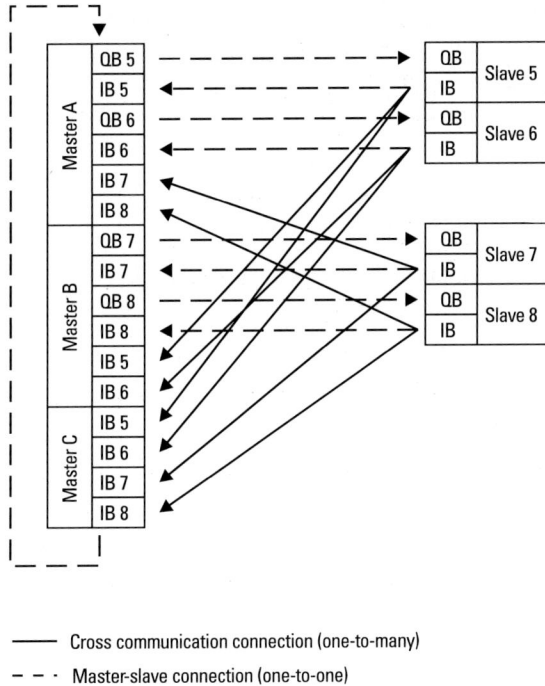

——— Cross communication connection (one-to-many)

– – · Master-slave connection (one-to-one)

Figure 2.9 Master-slave relationship during cross communication

2.4.2 Slave-Slave Relationship with Cross Communication

The slave-slave relationship shown in figure 2.10 with the use of I slaves (see section 3.4.3) such as the CPU315-2DP, is another version of data exchange by cross communication.

In this communication mode, the I slave is able to receive input data from other DP slaves.

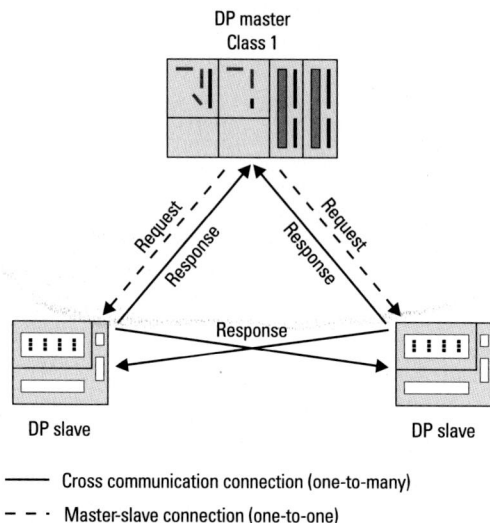

Figure 2.10   Slave-slave relationship during cross communication

2.5 DPV1 Functional Expansions

The increasingly complex demands on DP slaves require expanded communications functionality of PROFIBUS DP. This refers to the acyclic data traffic as well as to the interrupt functionality.

To meet these requirements, the international fieldbus standard EN 50170, Volume 2 has been expanded. The expansions described in the standard refer both to DP slave modules and to DP master modules. These function expansions, also known as DPV1 expansions, are optional to the standard protocol. This guarantees that previous PROFIBUS DP field devices and devices with DPV1 expansions can be operated together, thus ensuring inter-operability.

The following rules apply here:

- A DP slave with DPV1 expansion can be operated on a DP master without DPV1 functionality. The DPV1 functionality of the DP slave cannot be used.

- A DP slave without DPV1 expansion can be operated without restrictions on a DPV1 master.

A DP master with DPV1 expansions is also called a DPV1 master. This also applies for DP slaves that are designated DPV1 slaves with the expansions of EN 50170.

The expansion of the standard has thus cleared the way for a new generation of field devices. For planners and configuring engineers, however, the question of how exactly the various DP slave versions differ from each other usually remains.

- **DP standard slaves** have only the basic functionality described in the EN 50170 standard, that is, they have no DPV1 expansions. This means acyclic data traffic is not possible on a DP standard slave, and only the diagnostic interrupt is provided in the interrupt model. DP standard slaves are configured via a GSD file (device master file) in the relevant configuring tool.

- **S7 DP slaves** are further developed DP standard slaves from SIEMENS. However, these expansions can only be used with SIEMENS S7 DP master modules. Acyclic data traffic is possible with S7 DP slaves. An expanded interrupt model has also been implemented. If an S7 DP slave is configured using a GSD file and connected to DP master modules from other manufacturers, the S7 DP slave behaves like a DP standard slave without DPV1 expansions in accordance with EN 50170, Volume 2. The full functionality of S7 DP slaves is only achieved when configured with SIMATIC STEP 7 and operated on a SIMATIC S7 master module.

- **DPV1 slaves** are slaves with DPV1 expansions of EN 50170, Volume 2. These expansions refer to the interrupt model and the standardization of the acyclic data traffic. A DPV1 slave can be operated with full functionality on any DPV1 master. These slaves have a GSD file following revision level 3.

Table 2.2 gives an overview of which diagnostic or interrupt events can be initiated by which DP slave type. A requirement for this is that the slave is operated on an appropriate DP master.

Table 2.2 Availability of interrupts and of acyclic data traffic on DP slaves

	DP standard slave	S7-DP slave	DPV1 slave
Diagnostic interrupt	×	×	×
Process interrupt	–	×	×
Remove interrupt	–	×	×
Insert interrupt	–	×	×
Status interrupt	–	–	×
Update interrupt	–	–	×
Vendor-specific interrupt	–	–	×
Acyclic data traffic	No	Yes, with S7 DP master modules	Yes, with DPV1 master modules

3 PROFIBUS DP in SIMATIC S7 Systems

Introduction

PROFIBUS is an integral part of SIMATIC S7 systems. The I/O peripherals that are decentrally connected through the DP protocol are totally integrated in the system by the STEP 7 configuration tool. This means that already at the stage of configuration and programming, distributed I/O devices are treated in exactly the same way as the I/O connected locally in the central subrack or the expansion rack. The same applies to failure, diagnostics and alarm situations; the SIMATIC S7 DP slaves behave in the same way as I/O modules that are plugged in centrally. SIMATIC S7 provides integrated or plug-in PRO-FIBUS DP interfaces for the connection of field devices with more complex technical functions. Due to the properties of PROFIBUS Layer 1 and Layer 2 and the consistently implemented internal system communications (S7 functions), you can connect devices such as programming units (PG), PCs, and HMI and SCADA devices to a SIMATIC S7 PROFIBUS DP system.

3.1 PROFIBUS DP Interfaces in SIMATIC S7 Systems

We distinguish between two types of PROFIBUS DP interfaces in SIMATIC S7-300 and S7-400 systems:

▷ DP interfaces *integrated* on the CPUs (e. g. CPU 315-2DP, CPU 318-2DP, CPU 412-1, CPU 417-4)

▷ *Plug-in* DP interfaces through IM (*Interface Module*) or CP (*Communications Processor*) (IM 467, CP 443-5, and CP 342-5)

The performance data of the DP interfaces varies with the performance data of the CPUs. Tables 3.1 to 3.4 list the primary technical characteristics of both types of PROFIBUS DP interfaces – of those that are integrated into the CPUs, and of those that are plugged in the SIMATIC S7-300 and S7-400 systems. From the standpoint of configuration and program access, distributed I/O connected through DP interfaces is treated exactly in the same way as centralized I/O (with the exception of the CP 342-5). In contrast, the DP interface of the CP 342-5 operates independently of the CPU. The exchange of DP user data is managed by special function calls (FC) from within the user program.

In PROFIBUS DP systems, the S7-300-DP interfaces of CPUs CPU 315-2DP, and the CP 342-5 can be operated both as DP master and as DP slave. When you use the DP interface as a DP slave, you can select the mode for bus access control. Two modes are available: "DP slave as active node" and "DP slave as passive node." From the standpoint of the DP protocol, a DP slave that acts as an active node behaves as a (passive) DP slave during data exchange with the DP master. However, as soon as this "active DP slave" has the token, data can also be exchanged with other nodes due to additional communication services, such as FDL or S7 functions. This makes it possible to communicate with PGs, OPs and PCs and have data traffic from one S7 CPU to another through the DP interfaces of the SIMATIC S7 controllers, while the PROFIBUS DP functions are being executed.

Table 3.1 Technical data of the integrated PROFIBUS DP interfaces on S7-300 systems

Module	CPU 315-2DP		CPU 315-2DP		CPU 316-2DP	
MLFB number (order ref.)	6ES7 315-2AF01 6ES7 315-2AF02		6ES7 315-2AF03-0AB0		6ES7 316-2AG00-0AB0	
Number of interfaces	2 (1st interface for MPI only)		2 (1st interface for MPI only)		2 (1st interface for MPI only)	
Operating mode	DP master	DP slave	DP master	DP slave	DP master	DP slave
Baud rates in kbit/s	9.6 to 12,000	9.6 to 12,000	9.6 to 12,000	9.6 to 12,000	9.6 to 12,000	9.6 to 12,000
Max. number of DP slaves	64	–	64	–	64	–
Max. number of modules	512 total	32	512	32	512	32
Input bytes per slave	122 max.	–	244 max.	–	244 max.	–
Output bytes per slave	122 max.	–	244 max.	–	244 max.	–
Input bytes as slave	–	122 max.	–	244 max.	–	244 max.
Output bytes as slave	–	122 max.	–	244 max.	–	244 max.
Consistent data modules	32 bytes max.	32 bytes max.	32 bytes max.	32 bytes max.	32 bytes max.	32 bytes max.
Usable input area	1 kbyte		1 kbyte		2 kbytes	
Usable output area	1 kbyte		1 kbyte		2 kbytes	
Max. parameter data per slave	244 bytes		244 bytes		244 bytes	
Max. config. data per slave	244 bytes		244 bytes		244 bytes	
Max. diagnostics data per slave	240 bytes		240 bytes		240 bytes	
Cross communication support	No	No	Yes	Yes	Yes	Yes
Constant bus cycle time	No	–	Yes	–	Yes	–
SYNC/FREEZE	No	No	Yes	No	Yes	No
DPV1 mode	No	No	No	No	No	No

continue next page

Continuation of table 3.1

Module	CPU 318-2DP		
MLFB number (order ref.)	6ES7 318-2AJ00-0AB0		
Number of interfaces	2		
	1st interface	2nd interface	1st and 2nd interface
Operating mode	MPI / DP master	DP master / MPI	DP slave
Baud rates in kbit/s	9.6 to 12,000	9.6 to 12,000	9.6 to 12,000
Max. number of DP slaves	32	125	–
Max. number of modules	512	1024	32
Input bytes per slave	244 max.	244 max.	–
Output bytes per slave	244 max.	244 max.	–
Input bytes as slave	–	–	244 max.
Output bytes as slave	–	–	244 max.
Consistent data modules	–	–	32 bytes max.
Usable input area	2 kbytes	8 kbytes	–
Usable output area	2 kbytes	8 kbytes	–
Max. parameter data per slave	244 bytes	244 bytes	–
Max. config. data per slave	244 bytes	244 bytes	–
Max. diagnostics data per slave	240 bytes	240 bytes	–
Cross communication support	Yes	Yes	Yes
Constant bus cycle time	Yes	Yes	–
SYNC/FREEZE	Yes	Yes	No
DPV1 mode	From FW 3.0	From FW 3.0	From FW 3.0

Table 3.2 Technical characteristics of the plug-in PROFIBUS DP interfaces in S7-300 systems

Module	CP 342-5		CP 342-5	
MLFB number (order ref.)	6GK7 342-5DA00-0XA0 6GK7 342-5DA01-0XA0		6GK7 342-5DA02-0XA0	
Operating mode	DP master	DP slave	DP master	DP slave
Baud rates in kbit/s	9.6 to 1,500	9.6 to 1,500	9.6 to 12,000	9.6 to 12,000
Max. number of DP slaves	64	–	124	–
Max. number of modules	–	32	–	32
Input bytes per slave	240 max.	–	240 max.	–
Output bytes per slave	240 max.	–	240 max.	–
Input bytes as slave	–	86 max.	–	86 max.
Output bytes as slave	–	86 max.	–	86 max.
Consistent data modules	240 bytes max.	86 max.	240 bytes max.	86 max.
Usable input area	240 bytes max.	86 max.	240 bytes	86 max.
Usable output area	240 bytes max.	86 max.	240 bytes max.	86 max.
Max. parameter data per slave	242 bytes	–	242 bytes	–
Max. config. data per slave	242 bytes	–	242 bytes	–
Max. diagnostics data per slave	240 bytes	–	240 bytes	–
Cross communication support	No	No	No	No
Constant bus cycle time	No	No	No	No
SYNC/FREEZE	Yes	No	Yes	No
DPV1 mode	No	No	No	No

Table 3.3   Technical data of integrated PROFIBUS DP interfaces on S7-400 systems

Module	CPU 412-1	CPU 412-2		CPU 413-2
MLFB number (order ref.)	6ES7 412-1XF03-0AB0	6ES7 412-2XG00-0AB0		6ES7 413-2XG0?-0AB0
Number of interfaces	1	2		2 (1st interface for MPI only)
	1st interface	1st interface	2nd interface	2nd interface
Operating mode	MPI / DP master	MPI / DP master	DP master	DP master / MPI
Baud rates in kbit/s	9.6 to 12,000	9.6 to 12,000	9.6 to 12,000	9.6 to 12,000
Max. number of DP slaves	32	32	125	64
Input bytes per slave	244 max.	244 max.	244 max.	122 max.
Output bytes per slave	244 max.	244 max.	244 max.	122 max.
Consistent data modules	128 bytes max.	128 bytes max.	128 bytes max.	122 bytes max.
Usable input area	2 kbytes	2 kbytes	2 kbytes	2 kbytes
Usable output area	2 kbytes	2 kbytes	2 kbytes	2 kbytes
Max. parameter data per slave	244 bytes	244 bytes	244 bytes	244 bytes
Max. config. data per slave	244 bytes	244 bytes	244 bytes	244 bytes
Max. diagnostics data per slave	240 bytes	240 bytes	240 bytes	240 bytes
Cross communication support	Yes	Yes	Yes	No
Constant bus cycle time	Yes	Yes	Yes	No
SYNC/FREEZE	Yes	Yes	Yes	Only via ext. module (CP/IM)
DPV1 mode	From FW 3.0	From FW 3.0	From FW 3.0	No

continue next page

Continuation of table 3.3

Module	CPU 414-2	CPU 414-2		CPU 414-3	
MLFB number (order ref.)	6ES7 414-2X?00-0AB0 6ES7 414-2X?01-0AB0 6ES7 414-2X?02-0AB0	6ES7 414-2XG03-0AB0		6ES7 414-3XJ00-0AB0	
Number of interfaces	2 (1st interface for MPI only)	2		3 (3rd interface IF 964-DP can only be inserted as DP master)	
	2nd interface	1st interface	2nd interface	1st interface	2nd interface
Operating mode	DP master	MPI / DP master	DP master / MPI	MPI / DP master	DP master / MPI
Baud rates in kbit/s	9.6 to 12,000	9.6 to 12,000	9.6 to 12,000	9.6 to 12,000	9.6 to 12,000
Max. number of DP slaves	96	32	125	32	125
Input bytes per slave	122 max.	244 max.	244 max.	244 max.	244 max.
Output bytes per slave	122 max.	244 max.	244 max.	244 max.	244 max.
Consistent data modules	122 bytes max.	128 bytes max.	128 bytes max.	128 bytes max.	128 bytes max.
Usable input area	4 kbytes	2 kbytes	6 kbytes	2 kbytes	6 kbytes
Usable output area	4 kbytes	2 kbytes	6 kbytes	2 kbytes	6 kbytes
Max. parameter data per slave	244 bytes	244 bytes	244 bytes	244 bytes	244 bytes
Max. config. data per slave	244 bytes	244 bytes	244 bytes	244 bytes	244 bytes
Max. diagnostics data per slave	240 bytes	240 bytes	240 bytes	240 bytes	240 bytes
Cross communication support	No	Yes	Yes	Yes	Yes
Constant bus cycle time	No	Yes	Yes	Yes	Yes
SYNC/FREEZE	Only via ext. module (CP/IM)	Yes	Yes	Yes	Yes
DPV1 mode	No	From FW 3.0	From FW 3.0	From FW 3.0	From FW 3.0

continue page 54

Continuation of table 3.3

Module	CPU 416-2	CPU 416-2		CPU 416-3	
MLFB number (order ref.)	6ES7 416-2X?00-0AB0 6ES7 416-2X?01-0AB0	6ES7 416-2XK02-0AB0		6ES7 416-3XL00-0AB0	
Number of interfaces	2 (1st interface for MPI only)	2		3 (3rd interface IF 964-DP can only be inserted as DP master.)	
	2nd interface	1st interface	2nd interface	1st interface	2nd interface
Operating mode	DP master	MPI / DP master	DP master / MPI	MPI / DP master	DP master / MPI
Baud rates in kbit/s	9.6 to 12,000	9.6 to 12,000	9.6 to 12,000	9.6 to 12,000	9.6 to 12,000
Max. number of DP slaves	96	32	125	32	125
Input bytes per slave	122 max.	244 max.	244 max.	244 max.	244 max.
Output bytes per slave	122 max.	244 max.	244 max.	244 max.	244 max.
Consistent data modules	122 bytes max.	128 bytes max.	128 bytes max.	128 bytes max.	128 bytes max.
Usable input area	8 kbytes	2 kbytes	8 kbytes	2 kbytes	8 kbytes
Usable output area	8 kbytes	2 kbytes	8 kbytes	2 kbytes	8 kbytes
Max. parameter data per slave	244 bytes	244 bytes	244 bytes	244 bytes	244 bytes
Max. config. data per slave	244 bytes	244 bytes	244 bytes	244 bytes	244 bytes
Max. diagnostics data per slave	240 bytes	240 bytes	240 bytes	240 bytes	240 bytes
Cross communication support	No	Yes	Yes	Yes	Yes
Constant bus cycle time	No	Yes	Yes	Yes	Yes
SYNC/FREEZE	Only via ext. module (CP/IM)	Yes	Yes	Yes	Yes
DPV1 mode	No	From FW 3.0	From FW 3.0	From FW 3.0	From FW 3.0

continue next page

Continuation of table 3.3

Module	CPU 417-4		IF 964-DP
MLFB number (order ref.)	6ES7 417-4XL00-0AB0		6ES7 964-2AA00-0AB0
Number of interfaces	4 (3rd and 4th interface, IF 964-DP can only be inserted as DP master)		1
	1st interface	2nd interface	1st interface
Operating mode	MPI / DP master	DP master / MPI	On S7-400 CPUs, only DP master
Baud rates in kbit/s	9.6 to 12,000	9.6 to 12,000	9.6 to 12,000
Max. number of DP slaves	32	125	125 max. (for S7-400 CPUs)
Input bytes per slave	244 max.	244 max.	244 max. (for S7-400 CPUs)
Output bytes per slave	244 max.	244 max.	244 max. (for S7-400 CPUs)
Consistent data modules	128 bytes max.	128 bytes max.	128 bytes max. (for S7-400 CPUs)
Usable input area	2 kbytes	8 kbytes	Depends on CPU
Usable output area	2 kbytes	8 kbytes	Depends on CPU
Max. parameter data per slave	244 bytes	244 bytes	244 bytes (for S7-400 CPUs)
Max. config. data per slave	244 bytes	244 bytes	244 bytes (for S7-400 CPUs)
Max. diagnostics data per slave	240 bytes	240 bytes	240 bytes (for S7-400 CPUs)
Cross communication support	Yes	Yes	Depends on CPU
Constant bus cycle time	Yes	Yes	Depends on CPU
SYNC/FREEZE	Yes	Yes	Depends on CPU
DPV1 mode	From FW 3.0	From FW 3.0	Depends on CPU

Table 3.4   Technical data of the PROFIBUS DP plug-in interfaces in S7-400 systems

Module	IM 467 / IM 467-FO	IM 467	CP 443-5 Ext.	CP 443-5 Ext.
MLFB number (order ref.)	6ES7 467-5?J00-0AB0 6ES7 467-5?J01-0AB0	6ES7 467-5GJ02-0AB0	6GK7 443-5DX00-0XE0 6GK7 443-5DX01-0XE0	6GK7 443-5DX02-0XE0
Number of interfaces	1	1	1	1
Operating mode	DP master	DP master	DP master	DP master
Baud rates in kbit/s	9.6 to 12,000	9.6 to 12,000	9.6 to 12,000	9.6 to 12,000
Max. number of DP slaves	125	125	125	125
Input bytes per slave	244 max.	244 max.	244 max.	244 max.
Output bytes per slave	244 max.	244 max.	244 max.	244 max.
Consistent data modules	128 bytes max.	128 bytes max.	128 bytes max.	128 bytes max.
Usable input area	4 kbytes	4 kbytes	4 kbytes	4 kbytes
Usable output area	4 kbytes	4 kbytes	4 kbytes	4 kbytes
Max. parameter data per slave	244 bytes	244 bytes	244 bytes	244 bytes
Max. config. data per slave	244 bytes	244 bytes	244 bytes	244 bytes
Max. diagnostics data per slave	240 bytes	240 bytes	240 bytes	240 bytes
Cross communication support	No	Yes	No	Yes
Constant bus cycle time	No	Yes	No	Yes
SYNC/FREEZE	Yes	Yes	Yes	Yes
DPV1 mode	No	No	No	From 6GK7 443-5DX03-XE

3.2 Other Communication Functions Using DP Interfaces

In addition to the DP functions, the *active* DP interfaces (DP master and *active* DP slaves) of the SIMATIC S7-300 and S7-400 controllers support the following communication functions:

▷ S7 functions through integrated and plug-in DP interfaces

▷ PROFIBUS FDL services through communications processors (CPs) only

3.2.1 S7 Functions

The S7 functions offer communication services between CPUs of the S7 system, and to SIMATIC HMI systems (*H*uman *M*achine *I*nterface). All devices of the SIMATIC S7 series can handle the following S7 functions:

▷ Complete online functionality of STEP 7 for programming, testing, commissioning and diagnosing the SIMATIC S7-300/400 programmable controllers

▷ Read and write access to variables, and automatic transmission of data to HMI systems

▷ Transmission of data and data areas of 64 kbytes max. between individual SIMATIC S7 stations

▷ Read and write data between SIMATIC S7 stations, without any special communication user program on the communication partner

▷ Initiation of control functions such as STOP, warm and hot restart of the CPU of the communication partner

▷ Provision of monitoring functions, such as monitoring the operational status of a CPU of the communication partner

3.2.2 FDL Services

The FDL services provided by Layer 2 of PROFIBUS allow data blocks of up to 240 bytes to be sent and received. This type of communication is based on SDA telegrams (*S*end *D*ata with *A*cknowledge), and is used not only for the data traffic within SIMATIC S7 programmable controllers, but also for the data transfer between SIMATIC S7 and S5 systems and to PCs. In SIMATIC S7 controllers, the FDL services are handled by FUNC-TION CALLs (*AG_SEND* and *AG_RECV*) from within the user program.

3.3 System Response of the DP Interfaces in SIMATIC S7 Controllers

With the exception of the CP342-5, the DP master interfaces are totally integrated in the SIMATIC S7 concept as described in sections 3.3.1 to 3.3.8.

3.3.1 Startup Behavior of DP Master Interfaces in the SIMATIC S7

Particularly in plants with a distributed equipment layout, technical and topological factors often make it impossible to switch on all electrical machines or system parts at the same time. In practice, this may mean that not all DP slaves are available yet when the DP master starts up. Due to the time-staggered startup of power supplies and the resulting time-staggered startup of DP slaves, the DP master requires a certain startup time before it can load the slaves with the parameter sets, and start the cyclical exchange of user data with the DP slaves. For this reason, the S7-300 and S7-400 systems allow you to set the maximum delay for the READY message of all DP slaves after POWER-ON. The parameter "READY message from modules" sets this delay in the range from 1 to 65,000 milliseconds. The default value is 65,000 milliseconds. When the delay expires, the CPU goes to STOP or RUN, depending on the setting of the parameter "Startup for required configuration not equal actual configuration."

3.3.2 Failure/Recovery of DP Slave Stations

If a DP slave fails due to power failure, an interruption on the bus line or due to some other defect, the operating system of the CPU reports this error by calling organization block OB86 (failure of module rack, DP power or DP slave). OB86 is called for every type of event, irrespective of whether it is going or coming. If you don't program organization block OB86, the CPU will react to a DP power or slave failure by going into the STOP state. Thus, the SIMATIC S7 system reacts to faults in the distributed I/O modules in the same way as it would react to faults in the centralized I/O modules.

3.3.3 Insert/Remove Interrupt for DP Slave Stations

The insertion and removal of configured modules in SIMATIC S7 systems is monitored centrally. SIMATIC S7 DP slaves and DPV1 slaves can also monitor this event in distributed configurations and report to the DP master when it occurs. This starts organization block OB83 in the CPU, as a fault event when a module is removed and as a back-to-normal event when a module is inserted. When inserting a module into a configured slot in RUN mode, the CPU's operating system checks that the type of module inserted agrees with the configuration. OB83 is then started, and if the configured module type corresponds to the inserted module type, parameterization is carried out. If OB83 has not been programmed when an insert/remove interrupt occurs, the CPU changes to STOP mode.

3.3.4 Diagnostic Interrupts generated by DP Slave Stations

Distributed I/O modules with diagnostic capabilities are able to report events by generating a diagnostic interrupt. In this way, the DP slaves indicate error situations, such as partial node failure, wire break in signal modules, short-circuit or overload of an I/O channel, or failure of the load voltage supply. The CPU operating system reacts by calling organization block OB82 which is reserved for diagnostic interrupt processing. OB82, too, is called for every diagnostic interrupt, irrespective of whether this interrupt indicates a coming or a going event. If you don't program OB82, the CPU reacts to a diagnostic interrupt by going into the STOP state. Depending on the complexity of the DP slave, some of the possible diagnostic interrupts and their message formats are defined by the EN 50 170 standard. Others depend on the particular slave and manufacturer. With DP slaves from the SIMATIC S7 series, the diagnostic interrupts comply with the SIMATIC S7 system diagnostics.

3.3.5 Process Interrupts generated by DP Slave Stations

DP slaves of the SIMATIC S7 series and DPV1 slaves with process interrupt capability report process faults through the bus to the DP master (CPU). For example, a process interrupt would be generated if an analog input value were outside the defined limit range. In SIMATIC S7 systems, organization blocks OB40 to OB47 are reserved for process interrupts (also referred to as "hardware interrupts"). OB40 to OB47 are called by the operating system of the CPU when an interrupt occurs. If the relevant organization block has not been programmed, the CPU stays in RUN mode and does not change to STOP mode. Thus, the SIMATIC S7 CPU always reacts to process interrupts in the same way, irrespective of whether these interrupts are caused by centralized or distributed I/O modules. However, note that the reaction to process interrupts caused by distributed I/O is slower due to telegram run times on the bus and interrupt handling on the DP master.

3.3.6 Status Interrupt of DP Slave Stations

DPV1 slaves can trigger status interrupts. If, for example, a module of a DPV1 slave changes its operating status, from RUN mode, say, to STOP, this status change can be reported to the DP master using a status interrupt. The precise events that trigger a status interrupt are defined by the manufacturer and are detailed in the documentation of the DPV1 slave. A status interrupt causes the operating system of the CPU to call organization block OB55. If this OB has not been programmed, the CPU still remains in RUN. OB55 is only available on DPV1-enabled S7-CPUs.

3.3.7 Update Interrupt of DP Slave Stations

A DPV1 slave can, for example, signal the transfer of a parameter modification on a module by sending an update interrupt to the DP master. This causes OB56 to be called in the CPU. OB56 can only be programmed on DPV1-enabled S7-CPUs. When an update interrupt is received, the CPU always remains in RUN, even if OB56 has not been programmed. The manufacturer of the slave defines which event is signaled as an update interrupt by a DPV1 slave. Please refer to the description of the DPV1 slave for more detailed information.

3.3.8 Vendor-Specific Interrupt of DP Slave Stations

A vendor-specific interrupt can only be signaled to the DP master from a slot of a DPV1 slave. This causes the CPU to call organization block OB57. The organization block for vendor-specific interrupts is only available on DPV1-enabled S7-CPUs. If OB57 has not been programmed in the CPU, the CPU still remains in the RUN state. The manufacturer generally determines when a DPV1 slave triggers a vendor-specific interrupt, depending on the slave or, in the case of intelligent slaves, on their application. Please refer to the documentation of the slave for information on whether or when a DPV1 slave triggers a vendor-specific interrupt.

3.4 DP Slave Types in SIMATIC S7 Systems

SIMATIC S7 systems use three different groups of DP slaves. Depending on their configuration and their purpose, we classify the SIMATIC S7 DP slave devices in:

▷ *Compact* DP slaves

▷ *Modular* DP slaves

▷ *Intelligent* DP slaves (I slaves)

3.4.1 *Compact* DP Slaves

Compact DP slaves have a fixed structure in the input and output area which cannot be modified. The group of ET 200B electronic terminals (B stands for block I/O) is made up of such compact DP slaves. The ET 200B module series offers modules with different voltage ranges and different numbers of I/O channels.

3.4.2 *Modular* DP Slaves

With modular DP slaves, the structure of the input and output area is variable. You define it when you configure the DP slave using the S7 configuration software *HW Config*. ET 200M modules are typical representatives of this type of DP slave. You can connect up to eight I/O modules from the modular S7-300 series to one ET 200M interface module (IM 153).

3.4.3 *Intelligent Slaves* (I Slaves)

In a PROFIBUS DP network, S7-300 programmable controllers that contain a CPU of type CPU 315-2, CPU 316-2 or CPU 318-2, or the CP 342-5 communications processor, can be used as DP slaves. In SIMATIC S7 systems, these signal-conditioning field devices are referred to as "intelligent DP slaves;" I slaves for short. Use the S7 configuration software *HW Config* to define the structure of the input/output area for an S7-300 controller that acts as a DP slave.

One feature of intelligent DP slaves is that the input/output area provided to the DP master is not an actually existing I/O area. Instead, this input/output area is mapped by a preprocessing CPU.

4 Programming and Configuring PROFIBUS DP with STEP 7

STEP 7 is the standard programming and configuration software for SIMATIC S7 systems. This chapter describes the tools of the STEP 7 package (version 5.0, under Windows 95 or Windows NT) which you will use to set up and configure PROFIBUS DP networks. We assume that you have a STEP 7 software package installed on your PG programming unit or your PC, and that you are familiar with Windows 95 or Windows NT. The standard STEP 7 package comprises a number of applications (see figure 4.1), each of which does a specific job when programming an automation task, such as:

▷ Configuring the hardware and setting its parameters

▷ Configuring networks, connections and interfaces

▷ Creating and debugging user programs.

Additional optional software tools are available to expand the standard STEP 7 package for particular applications. These include programming language packages such as SCL, S7GRAPH or HiGraph. The graphic user interface provided for these tasks is known as *SIMATIC Manager*. *SIMATIC Manager* collects all the data and settings necessary for an automation task and combines this information into a project. Within this project, all data and settings are structured according to their function, and represented as objects. STEP 7 provides comprehensive online help including context-sensitive help for selected folders, objects and error messages.

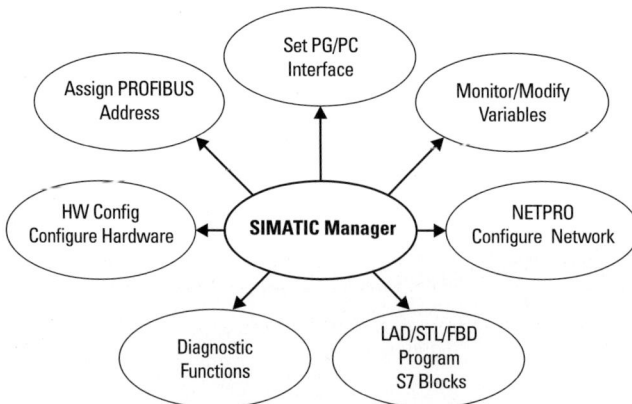

Figure 4.1
PROFIBUS DP-related STEP 7 applications which can be called from *SIMATIC Manager*

4.1 STEP 7 Fundamentals

4.1.1 STEP 7 Objects

In a similar way to the directory structure you know from *Windows Explorer* containing folders and files, a STEP 7 project is divided into folders and objects. Folders are objects which can contain other folders and objects. For example, the folder for an S7 station configured in *SIMATIC Manager* contains additional subfolders for the hardware and the S7 program. In turn, the S7 program contains additional folders for storage of text or graphic sources and STEP 7 software blocks making up the STEP 7 user program. The STEP 7 blocks which you create during project configuration and programming are stored in the *Blocks* folder in the form of objects.

Object orientation in STEP 7

When you process an object in *SIMATIC Manager*, the program automatically calls the application responsible for the type of object you are about to process. This automatic linking of objects to the related application makes processing of STEP 7 projects very easy. To start the application linked to the object, either double-click the object, or open the shortcut menu. To open the shortcut menu, select the object in *SIMATIC Manager*, and then click the right mouse button. In the shortcut menu, select OPEN OBJECT.

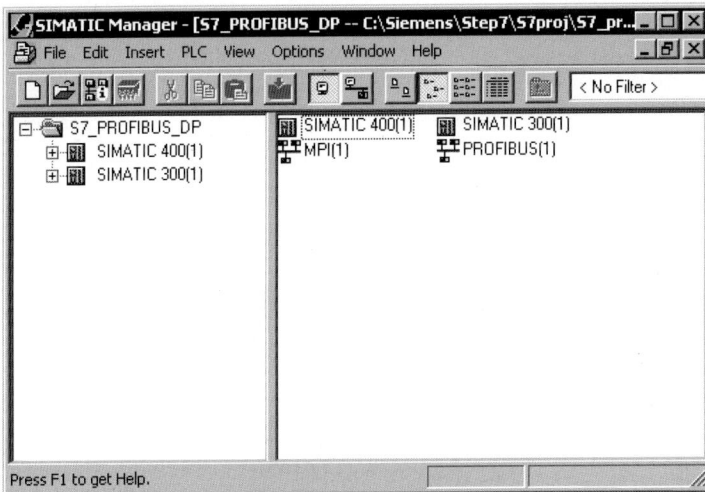

Figure 4.2 Folder and object structures in STEP 7 (example)

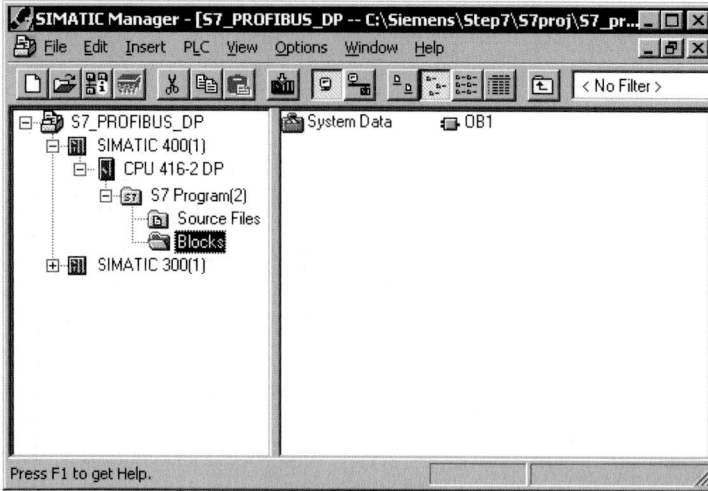

Figure 4.3 Object hierarchy in a STEP 7 project

4.1.2 STEP 7 Projects

The main object in *SIMATIC Manager* is the project. In the project, all data and programs required to handle an automation task are stored in a tree structure. This tree structure reflects the project hierarchy (see figure 4.3). The project is made up of the following configuration information:

▷ Configuration data on the hardware setup

▷ Parameter data for the modules used

▷ Configuration data for networks and communication

▷ Programs for programmable modules

4.2 Example Project with PROFIBUS DP

In this chapter we will develop an example project. While creating the project, we will explain the use of those STEP 7 programs that you need to set up and configure a SIMATIC S7 automation system using a PROFIBUS DP network. These are primarily the programs *SIMATIC Manager* and *HW Config*. The procedures suggested to create a SIMATIC S7 project will give you a simple and quick introduction to the STEP 7 configuration tool.

The example configuration is intended for a SIMATIC S7-400 programmable controller with a CPU416-2DP. We will connect the DP slaves ET200B-16DI/16DO, ET200M and S7-300/CPU315-2 through the integrated DP interface of the CPU, and set the transmission speed to 1500 kbit/s.

4.2.1 Creating a New STEP 7 Project

To create a new project, open *SIMATIC Manager*. Then go through the following procedure:

- In the menu bar, select FILE → New... to open the dialog box (figure 4.4) for setting up a new project.

- Select the "New project" button, and set the "Storage location (path)" for the new project.

- Enter a name (*S7-PROFIBUS DP* in our example) for the new project, and confirm and quit with OK.

You are now back in the main menu of *SIMATIC Manager*. The creation of the 7_PRO-FIBUS_DP object folder has automatically generated the *MPI* object (*Multi Point Interface*) which you can see in the right-hand half of the project screen. The *MPI* object is automatically generated by STEP 7 each time a new project is created. *MPI* is the standard programming and communication interface of the CPU.

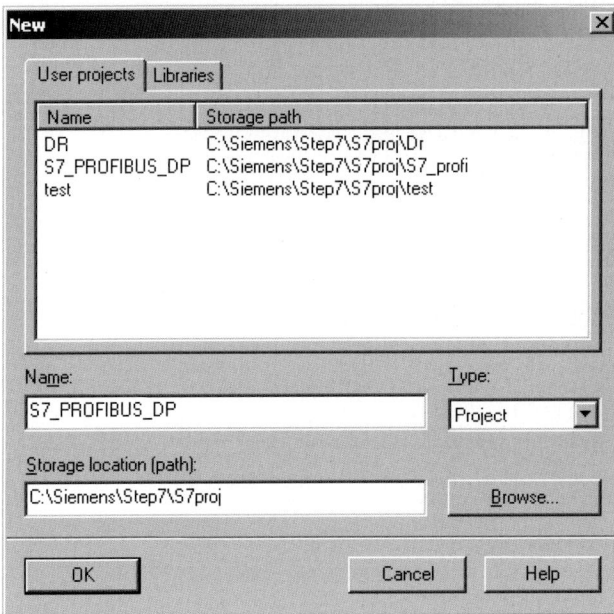

Name	Storage path
DR	C:\Siemens\Step7\S7proj\Dr
S7_PROFIBUS_DP	C:\Siemens\Step7\S7proj\S7_profi
test	C:\Siemens\Step7\S7proj\test

Name:
S7_PROFIBUS_DP

Type:
Project

Storage location (path):
C:\Siemens\Step7\S7proj

Browse...

OK   Cancel   Help

Figure 4.4
Dialog box for creating
a new project

4.2.2 Inserting Objects in the STEP 7 Project

In the left-hand side of the project screen, select the project. Open the shortcut menu using the right mouse button. Select the command INSERT NEW OBJECT, and insert the object SIMATIC 400 station. The newly inserted object appears in the right-hand half of the project screen. As with all other objects at this point you can change the object name, for example if you wish to give it a project-specific designation.

In the shortcut menu (remember: open with right click), select OBJECT PROPERTIES... In the Properties dialog box you could now enter some more characteristics for the object, such as Author's name, Comment, etc.

Next, insert the object *PROFIBUS* in the STEP 7 project which you have just created. Proceed as you did when inserting the SIMATIC 400 station.

4.2.3 PROFIBUS Network Settings

You are now back in the main project screen entitled S7-PROFIBUS DP. Select the object *PROFIBUS* and right-click to open the shortcut menu. Select OPEN OBJECT to call the graphic configuration tool NetPro. In the upper section of the screen, select the PROFIBUS subnet (PROFIBUS (1)), and right-click to open the shortcut menu. Select the command OBJECT PROPERTIES.... In the dialog box "Properties – PROFIBUS" open the "Network Settings" tab (see figure 4.5). Here you can set all relevant network parameters for the PROFIBUS subnet.

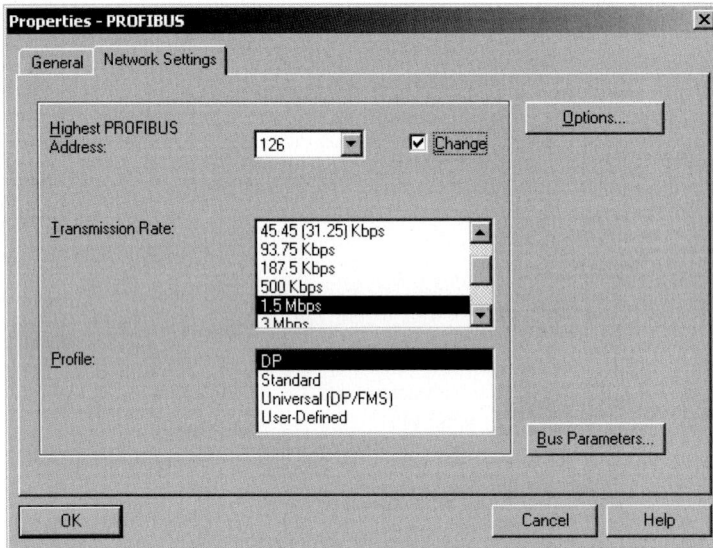

Figure 4.5   PROFIBUS network settings

Confirm the suggested settings (default settings) for the example project with OK. If you would like to begin creating the project immediately, skip to section 4.2.4.

In the following we shall briefly explain the meaning of the network parameters which you can set on the "Network Settings" tab of the "Properties – PROFIBUS" dialog box.

"Highest PROFIBUS address"

Referred to as HSA (*H*ighest-*S*tation-*A*ddress) in the EN 50170 standard. This parameter is used to optimize bus access control (token management) for multi-master bus configurations. With a mono-master PROFIBUS DP configuration, do not change the default setting of 126 for this parameter.

"Transmission Rate"

The transmission speed you select here will apply to the entire PROFIBUS subnet. This means that all stations (also called "nodes") which are used on this PROFIBUS subnet must support the selected baud rate. You can select baud rates from 9.6 to 12,000 kbit/s (Kbps). A baud rate of 1500 kbit/s is suggested as the default setting.

"Profile"

The bus profiles provide standards (default settings) for different PROFIBUS applications. Each bus profile contains a set of PROFIBUS bus parameters. These are calculated and set by the STEP 7 program, taking into account the particular configuration, profile and baud rate. These bus parameters apply globally to the entire bus and to all nodes connected to the PROFIBUS subnet.

You can define your own *"User-defined"* profile for your particular application. First, select the bus parameter settings of the profile *"DP"*, *"Standard"* or *"Universal (DP/ FMS)"*, save these as a user-defined profile, and then modify them as required. Such adjustments should of course only be made by engineers with networking experience.

Different bus profiles are available for different hardware configurations in your PROFIBUS DP network:

"DP" profile

Select this profile only if your system is a pure PROFIBUS DP mono-master and multi-master configuration containing SIMATIC S7 and SIMATIC M7 units. The optimized bus parameters calculated for this profile take into account all changes in the communication load when other nodes are connected to the bus. Such additional loads on the PROFIBUS subnet could be (one) PG programming unit, operator control and process monitoring devices, and configured non-cyclic FDL services, and FMS and S7 nodes.

The DP profile only considers those PROFIBUS nodes that are actually known to the PROFIBUS subnet. This means they must be part of the STEP 7 project and must have been properly configured.

"Standard" profile

Use this profile if you wish to extend the calculation of the bus parameters to other bus nodes which cannot be configured with STEP 7, or which do not belong to the currently processed STEP 7 project. On the "Network Settings" tab (figure 4.7), click the "Options..." button which opens the "Options" dialog box and the "Network Nodes" tab.

With the check box "Include network configuration below" disabled, the bus parameters are calculated with the same optimized algorithm used for the *"DP"* profile. If you enable this option, a simplified, more general algorithm is applied.

The *"Standard"* profile is especially designed for all other multi-master bus configurations (DP/FMS/FDL) with SIMATIC S7 and SIMATIC M7, and all configurations that extend over more than one STEP 7 project.

"Universal (DP/FMS)" profile

Use this bus profile if your network makes use of PROFIBUS components of the SIMATIC S5 series, such as the CP5431 communications processor or the S5-95U programmable controller. You should always select *"Universal (DP/FMS)"* when SIMATIC S7 and SIMATIC S5 stations are being used simultaneously on one PROFIBUS subnet.

Bus Parameters

The "Bus Parameters..." button provides access to the bus parameters calculated by STEP 7. Based on the bus configuration and the number of bus stations known in the STEP 7 project, STEP 7 calculates the values for the bus parameter "Ttr" (*Time Target Rotation*) and the bus parameter "Response monitoring" which is only relevant to PROFIBUS DP slaves.

Since the bus parameter "Ttr" (*Time Target Rotation*) calculated by STEP 7 represents a permissible maximum value and not the real token rotation time, you cannot use it to determine the reaction times of the bus system.

You can only change the values shown in figure 4.6 if you have selected the "User-defined" profile. Remember that a PROFIBUS subnet can only function reliably if the bus parameters are optimally tuned for the selected bus profile. The preset values displayed in the "Bus Parameters" dialog box should therefore only be changed by somebody with experience.

All bus parameter values are expressed in tBIT (time_BIT/run time). The bit run time shown in table 4.1 depends on the baud rate and is calculated as follows:

tBIT [μsec] = 1 / Mbit/s

Table 4.1 Bit run time as a function of the baud rate

Baud Rate	tBIT (μsec)
9.6 kbit/s	104.167
19.2 kbit/s	52.083
45.45 kbit/s	22.002
93.75 kbit/s	10.667
187.5 kbit/s	5.333
500 kbit/s	2.000
1,500 kbit/s	0.667
3,000 kbit/s	0.333
6,000 kbit/s	0.167
12,000 kbit/s	0.083

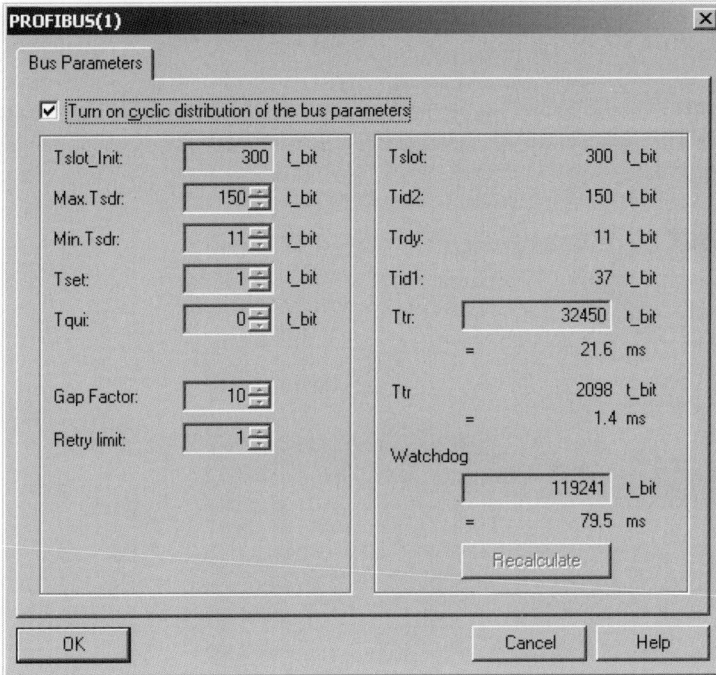

PROFIBUS(1)

Bus Parameters

☑ Turn on cyclic distribution of the bus parameters

Tslot_Init:	300	t_bit	Tslot:	300	t_bit
Max.Tsdr:	150	t_bit	Tid2:	150	t_bit
Min.Tsdr:	11	t_bit	Trdy:	11	t_bit
Tset:	1	t_bit	Tid1:	37	t_bit
Tqui:	0	t_bit	Ttr:	32450	t_bit
			=	21.6	ms
Gap Factor:	10		Ttr	2098	t_bit
			=	1.4	ms
Retry limit:	1		Watchdog		
				119241	t_bit
			=	79.5	ms

Recalculate

OK   Cancel   Help

Figure 4.6   Bus parameter settings

"Activate cyclic distribution of bus parameters"

When you enable this option, the parameter sets defined for the selected PROFIBUS subnet are transmitted cyclically by all DP master interfaces that are active in the PROFIBUS subnet. The data is transmitted in a multicast telegram issued by the SDN Service of Layer 2 (*Send Data* with *No* Acknowledge) with DSAP 63 (*Destination Service Access* Point).

Use this function if you wish to temporarily connect a PG programming unit to a running PROFIBUS subnet even though you don't know the exact bus parameter set for the PRO-FIBUS subnet. See also chapter 7.2 on setting the PG/PC interface.

You should not enable this function if you have selected the "constant bus cycle" mode (also referred to as "equidistant" bus cycle). This would increase the bus cycle unnecessarily. Neither should you enable it if the PROFIBUS subnet contains additional stations (third-party devices) which use DSAP 63 for multicast functions.

Option... "Constant Bus Cycle Time"

If you wish to run PROFIBUS DP in the Constant Bus Cycle Time mode (also referred to as "equidistant mode"), in the "Properties" dialog box, press the "Options" button and then open the "Constant Bus Cycle Time" tab (see figure 4.7). It describes the basic parameters for this mode. Only if the option "Constant Bus Cycle Time" is selected, are the other parameters in this dialog box enabled for selection. You can now set a constant bus cycle for the PROFIBUS subnet. A constant bus cycle means that the time interval between consecutive sending rights for the DP master is constant. Chapter 2.3.2 describes the principle of PROFIBUS DP operation with a constant bus cycle time (= equidistant bus cycle).

PROFIBUS subnets that are run at a constant bus cycle time may only contain one Class 1 DP master. Class 1 DP masters are DP masters which poll their DP slaves to exchange the I/O data cyclically.

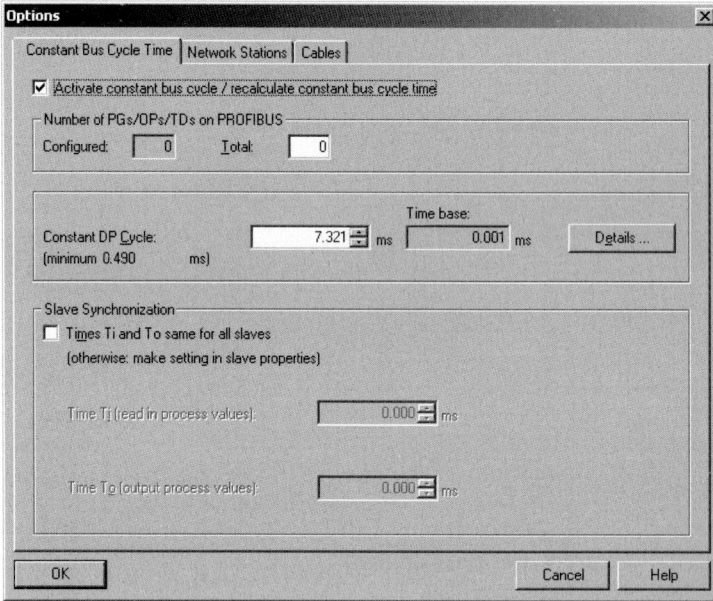

Figure 4.7   Setting the time values for the constant DP cycle (general values)

STEP 7 calculates and suggests a time value for the constant (= equidistant) DP cycle for the particular system configuration. This value is adequate to handle cyclic user data communication with the DP slaves and non-cyclic data exchange with PG, OP and TD devices within the suggested time. In the "Constant Bus Cycle Time" dialog box (see figure 4.7), you can reserve bus cycle time for additional PGs, OPs and TDs operated on the bus by setting the parameter "Number of PGs/OPs/TDs on the PROFIBUS" as required.

You may change the constant bus cycle time suggested by STEP 7. Increasing the suggested value is not a problem. However, if you wish to reduce the constant bus cycle (possibly even down to the displayed minimum value), you should remember that faults, such as DP slave failure and recovery, may have the effect of increasing the cycle to above the selected constant interval. Another adverse effect of reducing the constant interval to the minimum value is that the time available to other active nodes, such as PG programming units, for non-cyclic data exchange is also set to a minimum. In some networks, this can cause delays or even a breakdown of non-cyclic communication.

The "Details…" button takes you to another dialog box entitled "Constant Bus Cycle Time" (figure 4.8). It displays the time slices that make up the suggested constant bus cycle time. The time indicated for the cyclic portion is fixed and may not be changed. However, the non-cyclic time portion and the time portion available for additional active nodes, such as PG, OP and TD devices, can be modified.

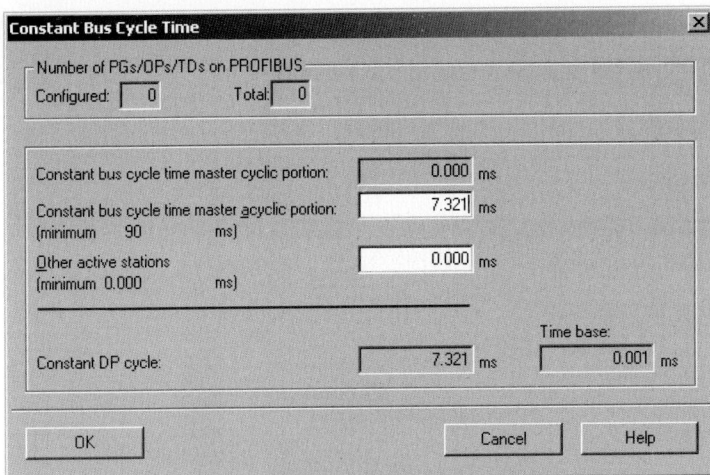

Figure 4.8 Setting the time values for the constant DP cycle (detailed values)

Options... "Network Nodes"

Your PROFIBUS system may contain nodes that cannot be registered by a STEP 7 project. To include such nodes in your bus system, select "Options..." in the "Properties – PROFIBUS" dialog box. The "Options" dialog box appears on the screen. Open the "Network Nodes" tab (figure 4.9). Here you can define how many additional active and passive nodes you wish to include in the bus parameter calculation. This option is not available for the "DP" profile.

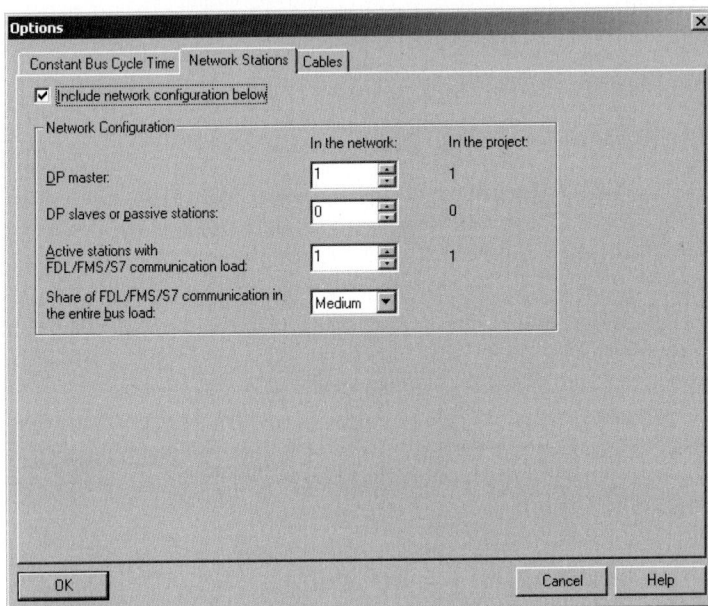

Figure 4.9 Additional network nodes for the PROFIBUS subnet

Options ... "Cables"

Not only the cable length, but also the use of RS 485 repeaters or OLMs (Optical Link Modules) with fiber optic cables affects the calculation of the bus parameters. The relevant variables are described on the "Cables" tab of the "Options" dialog box (figure 4.10).

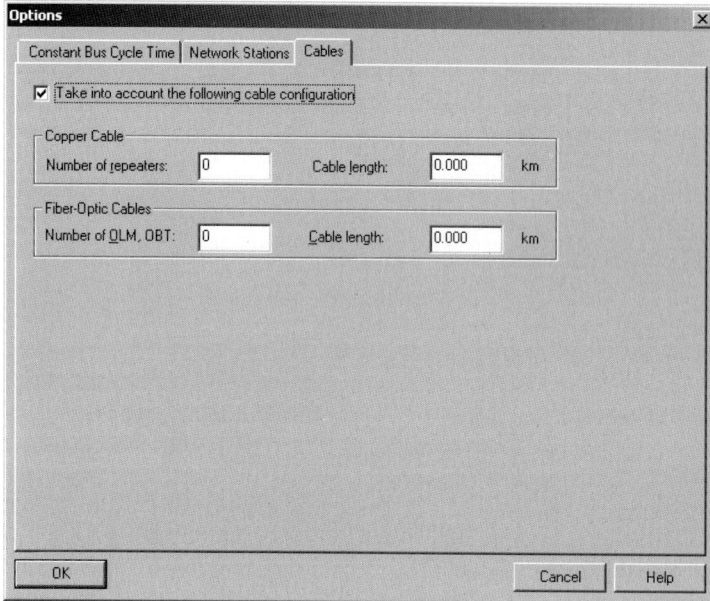

Figure 4.10 "Cables" dialog box for the definition of repeaters, OLMs and cable lengths

4.2.4 Configuring the Hardware Using the *HW Config* Program

The next step in setting up our PROFIBUS DP example network (see section 4.2.1) is the configuration of the hardware used on the S7-400 programmable controller. Quit the NetPro program, and return to the main screen of *SIMATIC Manager*. In the left-hand half, open the "S7_PROFIBUS_DP" folder with a double click. Then select the *SIMATIC 400(1)* object and call the *HW Config* program either by opening the shortcut menu with a right click and then selecting "OPEN OBJECT", or by double-clicking the hardware object in the right-hand half of the *SIMATIC Manager* screen. The *HW Config* program is automatically started, and a screen appears which is divided into two horizontal sections. At this stage it is still empty. Here you are going to configure the hardware for the SIMATIC 400 station.

Configuring the rack

In the toolbar, click the *Catalog* button, or in the menu bar select *View* → Catalog, to open the hardware catalog. In the catalog, open the *SIMATIC 400* folder. Under "RACK-400", select a rack. For our example configuration, select the UR 2 universal rack with 9 slots. Drag the selected rack to the upper left section of the screen.

The slots of the S7-400 rack are now listed in a configuration table. The lower part of the station screen displays detailed characteristics, such as order number, MPI address, and I/O addresses (I and Q).

Now, select the *PS407 10A* power supply from the hardware catalog *PS-400* and place it in slot 1 of the S7-400 rack. You will see that the selected power supply occupies two slots, slot 1 and slot 2.

Next, open the hardware catalog *CPU 400* → CPU 416-2DP and select the *CPU416-2DP* with order reference *6ES7 416-2XK00-0AB0*. Drag this CPU to slot 3 of the S7-400 rack. The *Parameters* tab of the *Properties – PROFIBUS Node DP Master* dialog box opens automatically. Here you can set the parameters of the DP master interface integrated on the CPU. Set the PROFIBUS address to "2," and in the table further down, select the PROFIBUS subnet (in our example, only one PROFIBUS subnet is configured) which you want to connect to the DP master interface of the CPU (see figure 4.11).

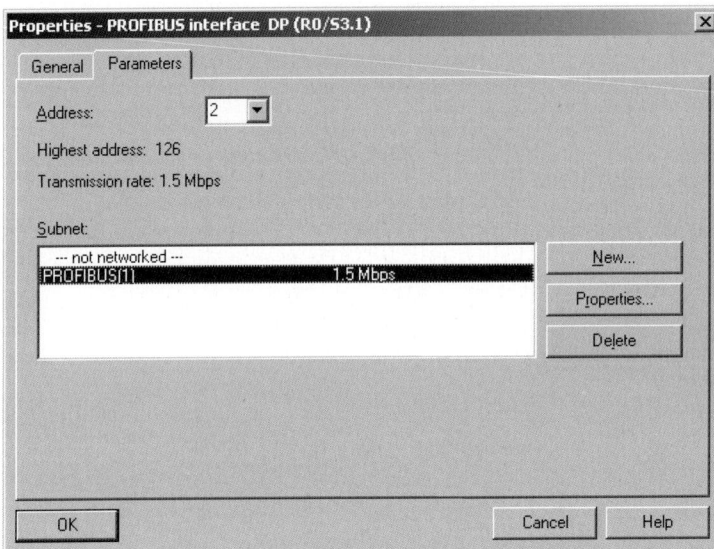

Figure 4.11   PROFIBUS network assignment "Properties – PROFIBUS Node DP Master"

In this dialog box, you can also set up a new PROFIBUS subnet or delete an already existing one.

Use OK to confirm your selections and return to the main screen of *HW Config*.

4.2.5 Configuring DP Slaves

Figure 4.12 shows the *HW Config* screen for the S7-400 station as it is configured up to now. The S7-400 station with the configured DP master system is displayed in the upper half of the screen.

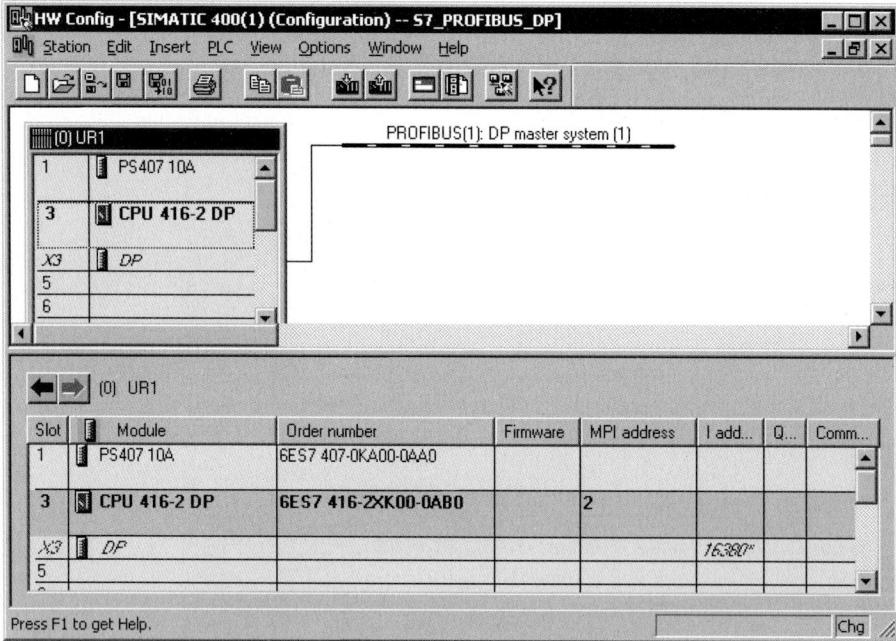

Figure 4.12   Station screen for the DP master system in *HW Config*

ET 200B station

In the next step, the DP slaves must be connected to the DP master system. To do this, open the *PROFIBUS DP* folder in the hardware catalog still displayed on the right-hand side of the screen. Open the *ET 200B* folder and select station *ET 200B-16DI/16DO*. Connect this DP slave to the DP master system by dragging it to the integrated DP Master interface shown in the upper left section of the screen. The *Properties – PROFIBUS Node B 16DI/16DO DO* dialog box automatically opens. Here, set the PROFIBUS address for this DP slave to "4," and click OK to return to the *HW Config* station screen.

The detailed view of the configured ET 200B station in the lower section of the screen (ET 200B station must be selected) indicates the addresses occupied by this DP slave (input bytes I "0" to "1" and output bytes Q "0" to "1") . See figure 4.13. If you want to change the addresses suggested by *HW Config* , double click the relevant line in the table. The "DP Slave Properties" dialog box opens and displays the actual structure of the input and output data. Here you could change the addresses, if required. In the startup phase, this data structure information is sent to the DP slave with the configuration telegram.

74

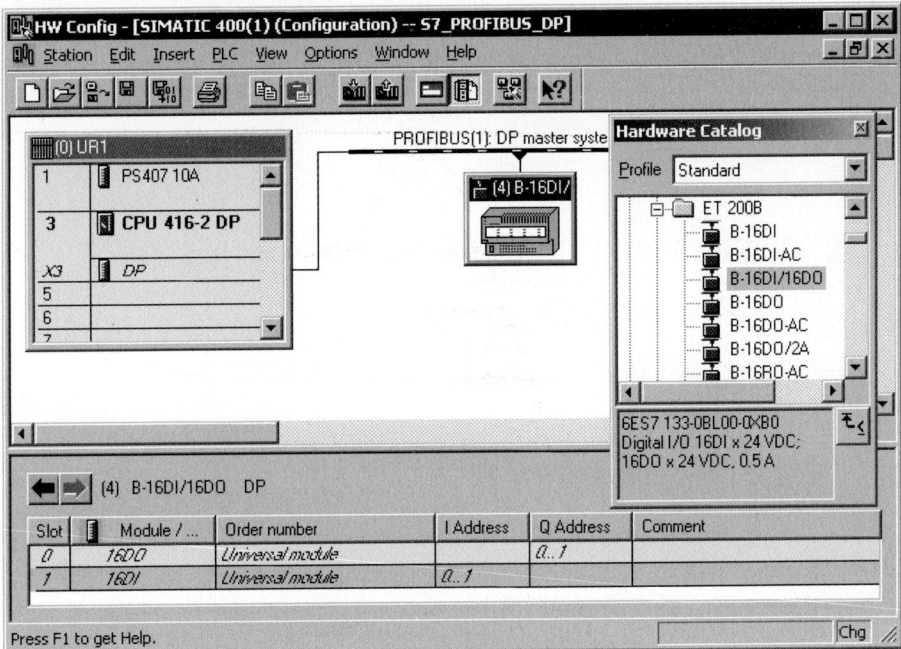

Figure 4.13   Station screen for the ET 200B DP slave *in HW Config*

To obtain a detailed view of the other station use the arrow button displayed in the top left corner to switch from one station to the other.

⬅ displays the detailed view of the selected DP slave (default setting)

➡ displays the detailed view of the DP master system.

Now, double-click the DP slave shown in the upper half of the *HW Config* station screen. This opens the *DP Slave Properties* dialog box and the *Properties* tab (figure 4.14). In *Properties* you see some reference information for the configured DP slave, such as order number, device family, type and description. Some other important characteristics must be set by you.

Diagnostic Address

The CPU uses this diagnostic address to indicate a DP slave failure with organization block OB86 ("Rack failure/DP slave failure"). In addition, you can read out diagnostic information from this address explaining the cause of the DP slave fault. The diagnostic address is suggested by *HW Config*. You can change it if necessary.

SYNC/FREEZE Capabilities

This field indicates whether the DP slave is capable of executing a SYNC and/or FREEZE control command issued by the DP master. *HW Config* gets this information from the GSD file (device master file) of the DP slave. At this point, the SYNC/FREEZE capabilities are indicated only. You cannot change the settings.

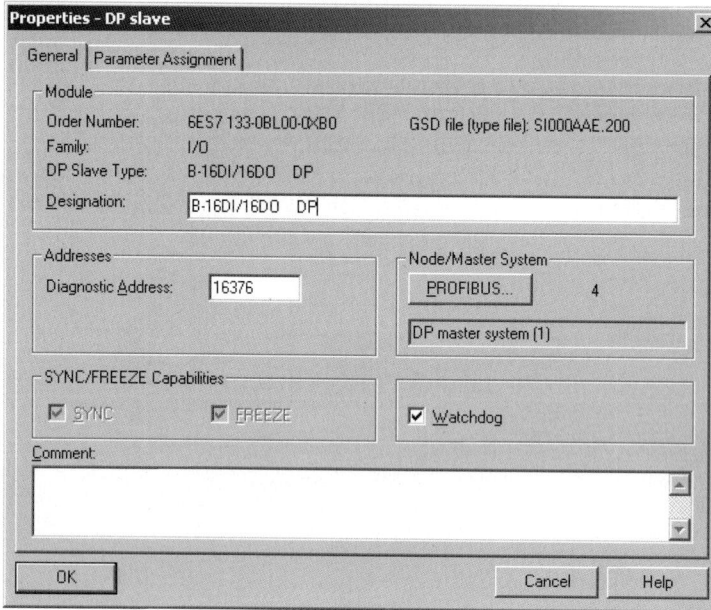

Figure 4.14 *Properties* dialog box of the DP slave

Response monitoring

Response monitoring should be switched on to allow the DP slave to react to a fault in the communication with the DP master. If there is no data communication between slave and master beyond the predefined response monitoring delay, the DP slave switches to a safe state. All outputs are set to signal status "0" or, if supported by the DP slave, to substitute values.

Note that hazardous system states may occur if response monitoring is disabled. Response monitoring can be switched on and off for every individual DP slave.

On the *Assigning Hexadecimal Parameters* tab of the *DP Slave Properties* dialog box, you can specify the slave-related parameters of the DP slave. For contents and meaning of this data, see the documentation of the DP slave device. For the ET 200B station configured in our example, it is not possible to set any of these hexadecimal parameters. However, you must always set 5 bytes of "00" (default setting). The information stored in this dialog box is transmitted to the DP slave as part of the parameter telegram.

For DP slaves of the SIMATIC S7 series, it is not necessary to specify any parameters in hexadecimal format. The settings for the parameter telegram are provided directly by the *HW Config* configuration tool during configuration of the DP slave.

ET 200M station

Our example configuration will also contain an ET 200M station. The ET200M device is of modular design and is equipped with an 8DI/8DO module, an AI2×12 bit module and an AO2×12 bit module. Follow the same configuration procedure as for the ET 200B station. In the hardware catalog, open the hardware *PROFIBUS DP* folder, then open the *ET 200M* folder and select the interface module *IM 153-2*. Connect the module to the S7-PROFIBUS DP network by dragging it to the integrated DP master interface. In the *Properties – PROFIBUS Node ET 200M IM 153-2* dialog box, set the PROFIBUS address for this DP slave to "5."

The detailed view of the configured ET 200M station displays a configuration table of eight lines which are numbered from 4 to 11. These 8 lines stand for the maximum of 8 modules from the S7-300 series which can be installed on the ET 200M station. To find the hardware modules of type *IM 153-2* that can be plugged into the ET 200 M unit, open the *IM 153-2* folder in the hardware catalog. The subfolders list the available modules. Open the *DI/DO-300* folder, select signal module *SM323 DI8/DO8×24 V/0.5 A* and move it to slot "4" of the detailed view of the ET 200M station, in the bottom section of the screen. Then, use the same procedure to place analog input module *SM 331 AI2×12 bits* in slot "5" and analog output module *SM 332 AO2×12 bits* in slot "6" of the ET 200M station (see figure 4.15).

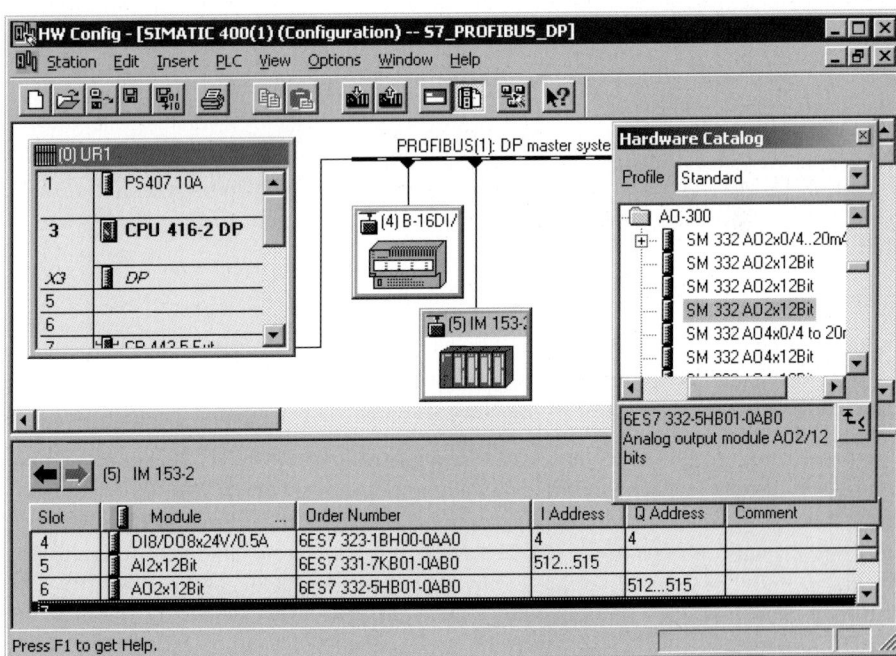

Figure 4.15 Station screen with detailed view of the ET 200M station in *HW Config*

Open the *Properties – AI2 × 12 bits* dialog box by double-clicking the analog input module *SM 331 AI2 × 12 bits* in line 5 of the detailed view. Open the *Inputs* tab to set the parameters of the analog inputs as required. The following settings are available.

- General enabling of interrupts
- Separate enabling of the diagnostic interrupts
- Enabling and setting the limit values for process interrupts
- Type of measurement
- Measuring range
- Position of the measuring range module
- Integration time

For our example, enable the diagnostic interrupts, and quit the *Inputs* tab with OK.

The following parameters can be set in the *Outputs* tab of the *Properties* dialog for analog output module *SM 332 AO2 × 12 bits* (double-click line 6 of the detailed view).

- Enable diagnostic interrupt
- Type of output
- Output range
- Reaction to CPU-STOP
- Substitute values, if applicable

In our example configuration, use the suggested default settings for the analog output module, and confirm and quit with OK.

The SIMATIC 400(1) master station is now complete. Save your settings with **Station →
Save and Compile** and then quit the SIMATIC 400(1) station screen with **Station →
Exit**.

S7-300/CPU315-2DP as I slave

Before you connect the S7-300 programmable controller to the DP master system, you have to set up this PLC (object) in the project. Proceed as described earlier, when we inserted the S7-400 station in the project (see section 4.2.2.)

To configure the modules for the S7-300 station, start with *SIMATIC Manager* and open the station screen for the S7-300 in *HW Config* (see also section 4.2.4). Open the hardware catalog, and select *SIMATIC 300* and *RACK-300*. Then select the object *Rail* and drag it to the upper section of the station screen. A configuration table indicating the slots of the S7-300 mounting rail appears. Place power supply *PS307 2A* from the "PS-300" hardware catalog in slot 1 of the module rack. Next, open the folders CPU-300 and CPU 315-2DP and select CPU315-2DP with the designation "6ES7 315-2AF01-0AB0." Move it to slot 2 of the module rack.

The *Properties – PROFIBUS Node DP Master* dialog box opens automatically. On the *Network Settings* tab, set the parameters for the DP interface integrated on the CPU. Set the PROFIBUS address to "6," and in the table further down, select the PROFIBUS subnet which you want to connect to the DP interface of the CPU. We will configure only one PROFIBUS subnet.

In our example, we will use the S7-300 programmable controller as a DP slave. Therefore, you must (re)configure the DP interface of the CPU315-2DP as a DP slave. To do this, double-click on the line *DPMaster* in the slot table. This opens the *Properties – DP Master* dialog box. On the *Operating Mode* tab, select the option "DP-Slave".

Now, change to the *Configuration* tab and select "new".

- Configuration of the input/output area on the DP slave for master-slave communication
- Configuration of the input/output area on the DP slave for direct data exchange (cross communication)
- Local diagnostic address of the DP slave interface. The diagnostic address on the *Addresses* tab is not relevant when the CPU is in slave mode.

Fill in the dialog box as shown in Figure 4.16. Click on "OK". The entered configuration is accepted as the module. Enter a second module in the same way but with Address type "Output", Address "1000", Length "10" and Consistency "All". Select "OK" to accept the values. The configuration shown in Figure 4.17 is then displayed.

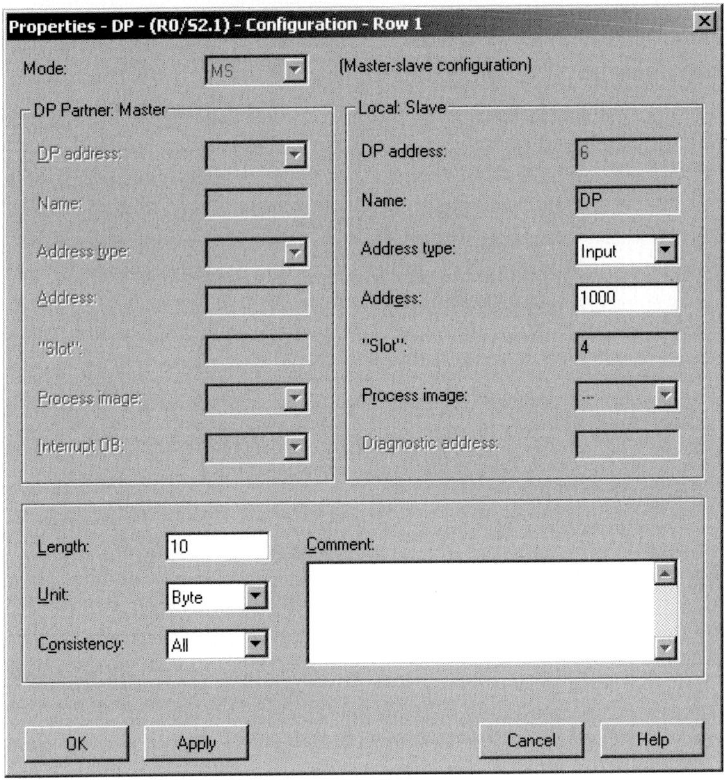

Figure 4.16 *HW Config,* "Properties of the DP Configuration"

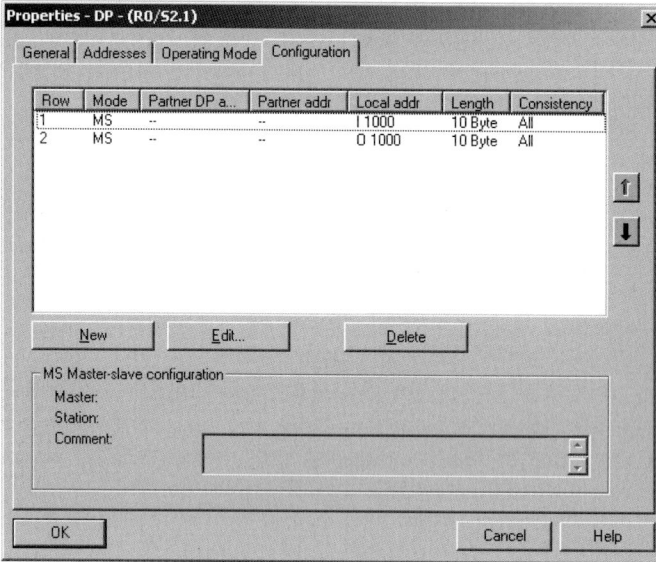

Figure 4.17 *HW Config, Properties – DP Interface*

Click OK to return to the *HW Config* main screen of the S7-300 station. The newly configured operating mode "DP-Slave" is now shown for the DP interface. Save the station
configuration for the S7-300 station in *HW Config*, and press the key combination "CTRL
+ TAB" to return to the S7-400 station screen. Finally, change to the "Configuration"
window and open the configuration dialog by double-clicking on the first line. Complete
the master-relevant parameters Address Type and Address specified in Figure 4.19, and
confirm with "OK". Now double-click to open the configuration of the second line, and
complete the Address Type with "Input" and Address with the value "1000". Click OK to
confirm the values and return to the "DP Slave Properties" window (Figure 4.20).

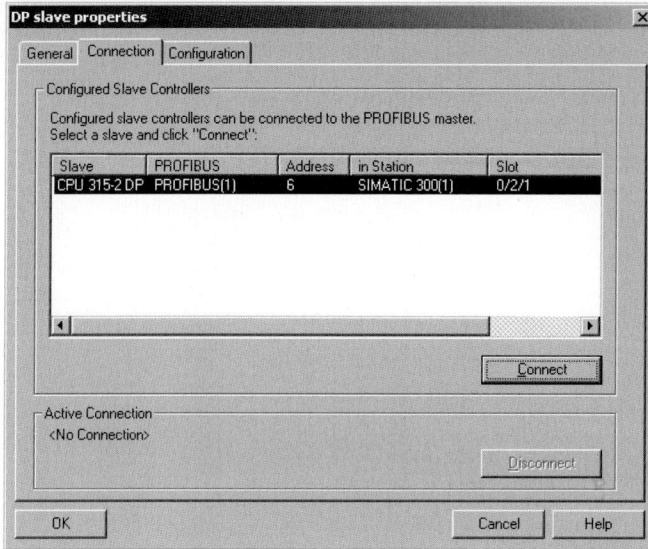

Figure 4.18 *HW Config*, *DP Slave Properties* dialog, *Connection* tab

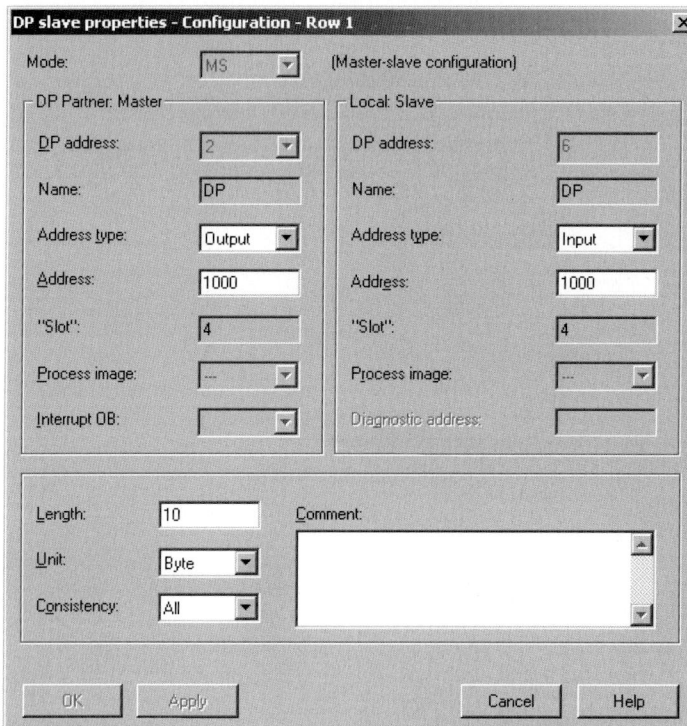

Figure 4.19 *HW Config*, *DP Master Properties* dialog, *Slave Configuration* tab

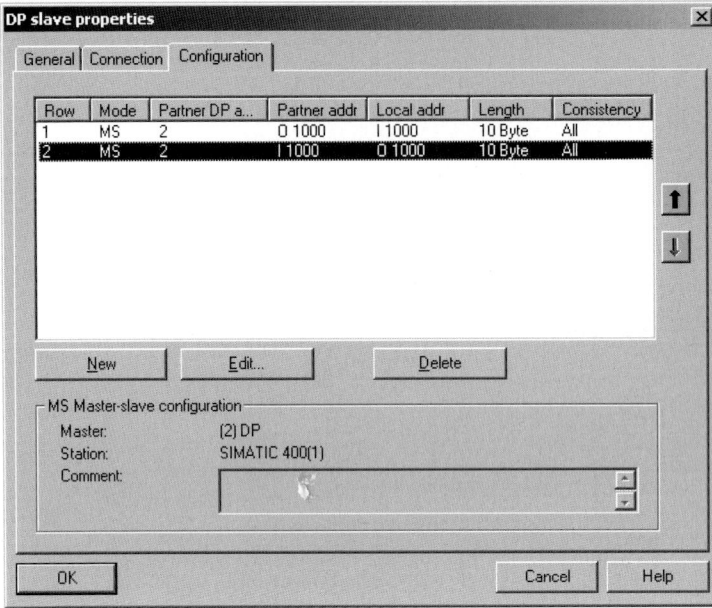

Figure 4.20 *HW Config,* "DP Slave Properties Configuration"

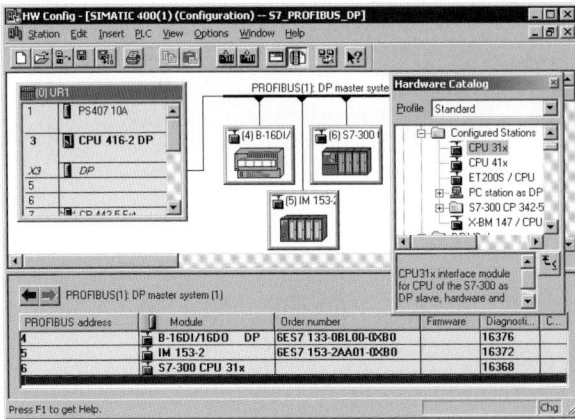

Figure 4.21 *HW Config,* station screen of the SIMATIC 400 station for sample project

5 DP User Program Interfaces

Introduction

The user program treats the distributed I/O connected to a SIMATIC S7 system in exactly the same way as those inputs and outputs that are locally connected in the central rack or expansion rack. Data communication with the DP slaves is handled by the process image input and output tables of the CPUs, or by direct I/O access commands from the user program.

Interfaces and functions are available to handle and evaluate process and diagnostic interrupts. The CPUs of the SIMATIC S7 series also allow you to change and adjust the parameter sets for the S7 DP and DPV1 slaves from within the user program.

A PROFIBUS network makes frequent use of coherent data areas with data structures using more than 4 bytes. Therefore, data communication with DP slaves that have more complex functions and data structures cannot be handled by simple I/O access commands from within the user program. SIMATIC S7 systems provide special system functions for communication with such DP slaves.

This chapter gives you an overview of general DP-related functions and interfaces of the SIMATIC S7 user program. The knowledge you gain from this chapter will help you to understand the practical examples described in chapters 6 and 7.

5.1 Basics of DP User Program Interfaces

5.1.1 Organization Blocks (OBs)

SIMATIC S7 CPUs offer a wide range of OBs (*Organization Blocks*) for executing the user program. Organization blocks are the interface between the CPU operating system and the user program. OBs are embedded in the user program. They have the task of executing special parts of the program upon certain events. For instance, when a hardware interrupt is triggered by an S7 DP or a DPV1 slave, or when a DP slave fails, the operating system of the S7 CPU calls the organization block dedicated to this situation. OBs permit event-controlled execution of the S7 user program. The event-driven call of an OB by the operating system usually interrupts the OB just being processed. Therefore, a system of priority classes (figure 5.1) applicable to all CPUs of the SIMATIC S7 series defines which OB is allowed to interrupt another OB. Higher priority OBs can interrupt lower priority OBs.

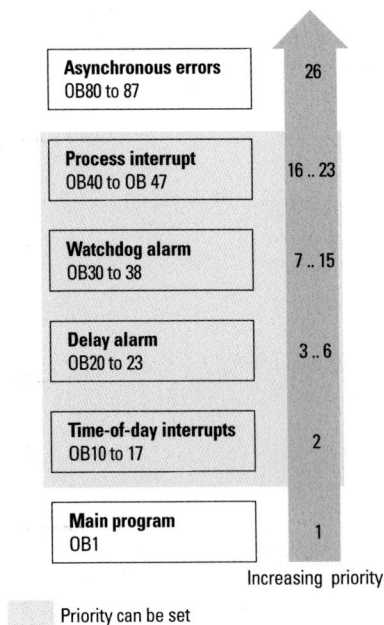

Figure 5.1
Priority classes of the
organization blocks

"1" is the lowest priority, and "29" is the highest priority.

When called by the operating system, each OB supplies 20 bytes of local data (variables) containing various pieces of information. The meaning of this local data depends on the OB. A description of the individual OBs and their local data is included in sections 5.2.1 to 5.2.7. The names of the variables are standard STEP 7 designations.

5.1.2 DP-Related System Functions (SFCs)

SIMATIC S7 systems provide a number of important DP-related system functions. These are called up by special functions integrated in the operating system of the S7 CPUs – the SFCs (*System Function Calls*).

General SFC parameters

Some SFC parameters are identical in meaning and function for all SFC calls described later in this chapter. This particularly applies to the SFC input parameters REQ, BUSY, LADDR, and the SFC output parameter RET_VAL.

REQ

Some SFCs have the REQ parameter to start the system function. If the REQ parameter transfers a logical "1" to the SFC when the function is called, the called function is executed. Remember that some SFCs are processed asynchronously (i.e., by means of several SFC calls and thus over several CPU cycles). Do not forget to consider the BUSY parameter.

BUSY

The BUSY parameter indicates whether the called SFC has been concluded. As long as the BUSY parameter is at "1," the called function is active.

LADDR

Depending on the SFC called, the LADDR input parameter defines either the logical start address configured in the *HW Config* program or the diagnostic address of the DP slave resp. of the input/output module. Although this address is configured in the *HW Config* program in decimal format, it must be specified in hexadecimal format in the block.

RET_VAL

All SFCs have an output parameter called RET_VAL. It returns a value informing whether execution of the system function was successful. If an error occurs while the SFC is being processed, the return value contains an error code. The return values in the RET_VAL output parameter of SFCs are divided into two classes.

▷ General error codes
▷ SFC-related error codes

The return value (figure 5.2) shows whether the error code is general or specific.

The general error codes described in table 5.1 are the same for all system functions. The format of the error code is hexadecimal. The letter "x" in the error code numbers is only used as a placeholder. It stands for the number of the parameter of the system function that caused the error.

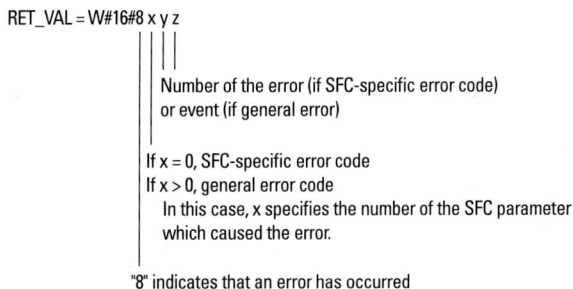

RET_VAL = W#16#8 x y z

> Number of the error (if SFC-specific error code)
> or event (if general error)

> If x = 0, SFC-specific error code
> If x > 0, general error code
> In this case, x specifies the number of the SFC parameter
> which caused the error.

> "8" indicates that an error has occurred

Figure 5.2   Layout of the RET_VAL SFC parameter

Table 5.1   General RET_VAL error codes

Error Code w#16#....	Explanation
8x7F	Internal error
	This error code indicates an internal error in parameter x. This error was not caused by the user and can therefore not be corrected by the user.
8x22	Area length error while reading a parameter
8x23	Area length error while writing a parameter
	This error code indicates that parameter x is completely or partially outside the address range, or the length of a bit field for a parameter of data type "ANY-Pointer" is not divisible by 8.
8x24	Area length error while reading a parameter
8x25	Area length error while writing a parameter
	This error code indicates that parameter x is located in an area which is not permitted for the system function. The description of the function concerned indicates the areas that are invalid for this function.
8x28	Offset error while reading a parameter
8x29	Offset error while writing a parameter
	This error code indicates that the reference to parameter x is an address whose bit address is not "0".
8x30	The parameter is located in the write-protected global DB.
8x31	The parameter is located in the write-protected instance DB. This error code indicates that parameter x is located in a write-protected data block. If the data block was opened by the system function itself, the system function always provides the value W#16#8x30 at its output.
8x32	The parameter contains a DB number which is too large (number error of the DB).
8x34	The parameter contains an FC number which is too large (number error of the FC).
8x35	The parameter contains an FB number which is too large (number error of the FB).
	This error code indicates that parameter x contains a block number which is larger than the maximum permissible block number.
8x3A	The parameter contains the number of a DB which is not loaded.
8x3C	The parameter contains the number of an FC which is not loaded.
8x3E	The parameter contains the number of an FB which is not loaded.
8x42	An access error occurred while the system was attempting to read a parameter from the I/O area of the inputs.
8x43	An access error occurred while the system was attempting to write a parameter to the I/O area of the outputs.
8x44	Error during n^{th} (n > 1) read access after occurrence of an error
8x45	Error during n^{th} (n > 1) write access after occurrence of an error
	This error code indicates that access to the requested parameter is not permitted.

Memory areas used for the SFC call parameters

For the identifiers used for the memory areas with the SFC parameters, see table 5.2.

Table 5.2 Memory areas of the SFC parameters

Type	Memory Area	Unit
E	Process image input table	Input (bit)
		Input byte (IB)
		Input word (IW)
		Input double word (ID)
A	Process image output table	Output (bit)
		Output byte (QB)
		Output word (QW)
		Output double word (QD)
M	Memory	Bit memory (bit)
		Memory byte (MB)
		Memory word (MW)
		Memory double word (MD)
D	Data block	Data bit
		Data byte (DBB)
		Data word (DBW)
		Data double word (DBD)
L	Local data	Local data bit
		Local data byte (LB)
		Local data word (LW)
		Local data double word (LD)

5.1.3 Basics of SIMATIC S7 Data Records

All system data and parameters are stored on S7 modules in the form of data records. The individual data records are numbered from 0 to a maximum of 240. Not all modules have access to all data records.

Depending on the type of S7 module used, there are system data areas which the user program can only write to, or only read from.

Table 5.3 shows the layout of the system data areas which can only be written to. It lists how large the individual data records may be and which SFCs can be used to transfer them to the modules.

Table 5.4 shows the layout of the system data areas which can only be read. The table lists how large the individual data records may be and which SFCs can be used to read them.

Table 5.3 Layout of system data areas on S7 modules which can only be written to

Data Record Number	Contents	Size	Restrictions	Write Access With
0	Parameter	With S7-300: 2 to 14 bytes	Write access for S7-400 only	SFC56 *WR_DPARM* SFC57 *PARM_MOD*
1	Parameter	With S7-300: 2 to 14 bytes (DR0 and DR1 have exactly 16 bytes together.)	–	SFC55 *WR_PARM* SFC56 *WR_DPARM* SFC57 *PARM_MOD*
2 to 127	User data	Up to 240 bytes	–	SFC55 *WR_PARM* SFC56 *WR_DPARM* SFC57 *PARM_MOD* SFC58 *WR_REC*
128 to 240	Parameter	Up to 240 bytes	–	SFC55 *WR_PARM* SFC56 *WR_DPARM* SFC57 *PARM_MOD* SFC58 *WR_REC*

Table 5.4 Layout of the system data areas on modules which can only be read

Data Record Number	Contents	Size	Read Access With
0	Module-related diagnostic data	4 bytes	SFC51 *RDSYSST* (INDEX 00B1H) SFC59 *RD_REC*
1	Channel-related diagnostic data (incl. data record 0)	– With S7-300: 16 bytes – With S7-400: 7 to 220 bytes	SFC51 *RDSYSST* (INDEX 00B2H and 00B3H) SFC59 *RD_REC*
2 to 127	User data	Up to 240 bytes	SFC59 *RD_REC*
128 to 240	Diagnostic data	Up to 240 bytes	SFC59 *RD_REC*

For every newly triggered transmission of a data record, resources (i.e., memory space) are allocated on the CPU for the asynchronous job processing of the SFCs. When several jobs are active at the same time, it is ensured that all jobs will be performed and that they do not affect each other. However, only a certain number of SFC calls may be active simultaneously. For the maximum possible number of SFC calls, see the performance data of the S7 CPU you are using. When the maximum amount of allocable resources is reached, an appropriate error code is transmitted by the RET_VAL parameter. If this happens, the SFC must be triggered again.

The individual parameters in the data records can be static or dynamic. To change a static module parameter, such as an input delay of a digital input module, you must use the HW Config program.

Dynamic parameters are handled differently. They can be changed during running operation. For instance, if you want to change the limit values of an analog input module, you can do that by calling an SFC.

5.2 Organization Blocks

5.2.1 Cyclically Processed Main Program (OB1)

The main program is executed in OB1. OB1 calls up function blocks (FBs), standard function blocks (SFBs) or functions by means of function calls (FCs) and system function calls (SFCs). OB1 is processed cyclically. It is executed for the first time after the startup OBs have been processed (OB100 for warm restart or OB101 for hot restart or OB102 for cold restart). At the end of the OB1 cycle, the operating system transfers the process image output table to the output modules. Before OB1 is started again, the operating system updates the process image input table by reading the current signal states of the input I/O. This procedure is continuously repeated. This is what we call "cyclic processing." Of all the OBs whose run-time is monitored, OB1 has the lowest priority. It can therefore be interrupted by higher priority OBs.

The CPUs of SIMATIC S7 programmable controllers allow you to monitor the maximum cycle time. This is the processing time for OB1. You can also ensure that a minimum cycle time for processing OB1 is observed. If you have set a minimum cycle time, the CPU operating system delays another start of OB1 until this time is reached. You can define the parameters for cycle monitoring and minimum cycle time in the *HW Config* program, under *CPU Properties*. For the meaning of the local data of OBs, see table 5.5.

Table 5.5 Local data of OB1

Variable	Data Type	Description
OB1_EV_CLASS	BYTE	Event class and identifiers: B#16#11 = Alarm is active
OB1_SCAN_1	BYTE	B#16#01 = Conclusion of warm restart B#16#02 = Conclusion of hot restart B#16#03 = Conclusion of free cycle
OB1_PRIORITY	BYTE	Priority class "1"
OB1_OB_NUMBR	BYTE	OB number (01)
OB1_RESERVED_1	BYTE	Reserved
OB1_RESERVED_2	BYTE	Reserved
OB1_PREV_CYCLE	INT	Run time of the previous cycle (msec)
OB1_MIN_CYCLE	INT	Minimum cycle time (msec) since the last startup
OB1_MAX_CYCLE	INT	Maximum cycle time (msec) since the last startup
OB1_DATE_TIME	DT	Date and time at which the OB was requested

Format for the representation of hexadecimal numbers

Data type Byte   B#16#x   (Value range for x from "0" to "FF")
Data type Word   W#16#x   (Value range for x from "0" to "FFFF")
Data type Double Word DW#16#x   (Value range for x from "0" to "FFFFFFFF")

5.2.2 Process Interrupts (OB40 to OB47)

The CPUs of the SIMATIC S7 programmable controllers provide eight different OBs (OB40 to OB47) for reacting to process interrupts. For those S7 DP slaves that support process interrupts, you can define the channels, general conditions and the OB number in the *HW Config* hardware configuration program.

When an S7 DP slave triggers a process interrupt, this is identified by the CPU operating system, and the appropriate OB is started according to its priority. After the user program in the interrupt OB has been processed (the OB is concluded), an acknowledgment message is sent to the S7 DP slave that caused the interrupt.

If another interrupt signal arrives while the OB is still processing the first interrupt, the second request is registered and stored, and the OB is processed at the appropriate time. This is the procedure in S7-400 systems. With S7-300 systems, the second process interrupt is lost if the event causing it is no longer queued after acknowledgment of the interrupt just processed.

The process interrupt OBs supply 20 bytes of local data. Among other things, this data contains the logical base address of the module which generated the interrupt. For a description of the local data, refer to table 5.6.

Table 5.6   Contents of the local data, OB40 to OB47

Variable	Data Type	Description
OB4x_EV_CLASS	BYTE	Event class and identifiers: B#16#11 = interrupt is active
OB4x_STRT_INF	BYTE	B#16#41 = interrupt through interrupt line 1 Only with S7-400: B#16#42 = Interrupt through interrupt line 2 B#16#43 = Interrupt through interrupt line 3 B#16#44 = Interrupt through interrupt line 4
OB4x_PRIORITY	BYTE	Priority class "16" (OB40) to "23" (OB47) (default)
OB4x_OB_NUMBR	BYTE	OB no. (40 to 47)
OB4x_RESERVED_1	BYTE	Reserved
OB4x_IO_FLAG	BYTE	B#16#54 = input module B#16#55 = output module
OB4x_MDL_ADDR	WORD	Logical base address of the module which triggered the interrupt
OB4x_POINT_ADDR	DWORD	For digital modules: Bit field with the states of the inputs on the module For analog modules (CP or IM): Interrupt state of the module
OB4x_DATE_TIME	DT	Date and time at which the OB was requested

5.2.3 Status Interrupt (OB55)

OB85 is called by the operating system of the S7-CPU when a status interrupt has been triggered from a slot of a DPV1 slave. The status interrupt OB is only available on DPV1-enabled S7-CPUs. These CPUs provide the OB55 for processing status interrupts. If OB55 has not been programmed, the CPU remains in RUN, and an entry is made in the diagnostic buffer of the CPU.

The status interrupt OB supplies 20 bytes of local data. Among other things, this data contains the logical base address and the slot of the module that generated the status interrupt. Please refer to Table 5.7 for a description of the local data.

Table 5.7   Local data of OB55

OB55_EV_CLASS	BYTE	Event class and identifiers: B#16#11 (coming interrupt)
OB55_STRT_INF	BYTE	B#16#55 (start request for OB55)
OB55_PRIORITY	BYTE	Parameterized priority class, default value: 2
OB55_OB_NUMBR	BYTE	OB number (55)
OB55_RESERVED_1	BYTE	Reserved
OB55_IO_FLAG	BYTE	Input module: B#16#54 Output module: B#16#55
OB55_MDL_ADDR	WORD	Logical base address of the interrupt-generating component (module)
OB55_LEN	BYTE	Length of the data block supplying the interrupt
OB55_TYPE	BYTE	Identifier for the interrupt type "Status interrupt"
OB55_SLOT	BYTE	Slot number of the interrupt-generating component (module)
OB55_SPEC	BYTE	Specifier: Bit 0 to 1: Interrupt specifier Bit 2: Add_Ack Bit 3 to 7: Sequence number
OB55_DATE_TIME	DT	Date and time at which the OB was requested

5.2.4 Update Interrupt (OB56)

SIMATIC S7-CPUs recognize and register the arrival of an update interrupt by calling OB56. The update interrupt OB is only available on DPV1-enabled S7-CPUs. It is called by the operating system of the CPU if an update interrupt has been generated from a slot or a module of a DPV1 slave. If OB56 has not been programmed, the CPU remains in RUN and an entry is made in the diagnostic buffer of the CPU.

The update interrupt OB supplies 20 bytes of local data. Among other things, this data contains the logical base address and the slot of the module that generated the status interrupt. Please refer to Table 5.8 for a description of the local data.

Table 5.8   Local data of OB56

OB56_EV_CLASS	BYTE	Event class and identifiers: B#16#11 (coming interrupt)
OB56_STRT_INF	BYTE	B#16#56 (start request for OB56)
OB56_PRIORITY	BYTE	Parameterized priority class, default value: 2
OB56_OB_NUMBR	BYTE	OB number (56)
OB56_RESERVED_1	BYTE	Reserved
OB56_IO_FLAG	BYTE	Input module: B#16#54 Output module: B#16#55
OB56_MDL_ADDR	WORD	Logical base address of the interrupt-generating component (module)
OB56_LEN	BYTE	Length of the data block supplying the interrupt
OB56_TYPE	BYTE	Identifier for the interrupt type "Update interrupt"
OB56_SLOT	BYTE	Slot number of the interrupt-generating component (module)
OB56_SPEC	BYTE	Specifier: Bit 0 to 1: Interrupt specifier Bit 2: Add_Ack Bit 3 to 7: Sequence number
OB56_DATE_TIME	DT	Date and time at which the OB was requested

5.2.5 Vendor-Specific Interrupt (OB57)

The SIMATIC S7-CPUs provide OB57 for detecting vendor-specific interrupts. The operating system of the CPU calls OB57 if a vendor-specific interrupt has been generated from a slot of a DPV1 slave. The interrupt OB for vendor-specific interrupts is only available on DPV1-enabled S7-CPUs. If OB57 has not been programmed, the CPU remains in RUN. An entry is simply made in the diagnostic buffer of the CPU.

OB 57 supplies 20 bytes of local data. Among other things, this data contains the logical base address and the slot of the module that generated the interrupt. Please refer to Table 5.9 for a description of the local data.

Table 5.9 Local data of OB57

OB57_EV_CLASS	BYTE	Event class and identifiers: B#16#11 (coming interrupt)
OB57_STRT_INF	BYTE	B#16#57 (start request for OB57)
OB57_PRIORITY	BYTE	Parameterized priority class, default value: 2
OB57_OB_NUMBR	BYTE	OB number (57)
OB57_RESERVED_1	BYTE	Reserved
OB57_IO_FLAG	BYTE	Input module: B#16#54 Output module: B#16#55
OB57_MDL_ADDR	WORD	Logical base address of the interrupt-generating component (module)
OB57_LEN	BYTE	Length of the data block supplying the interrupt
OB57_TYPE	BYTE	Identifier for the interrupt type "Update interrupt"
OB57_SLOT	BYTE	Slot number of the interrupt-generating component (module)
OB57_SPEC	BYTE	Specifier: Bit 0 to 1: Interrupt specifier Bit 2: Add_Ack Bit 3 to 7: Sequence number
OB57_DATE_TIME	DT	Date and time at which the OB was requested

5.2.6 Diagnostic Interrupts (OB82)

The CPUs of SIMATIC S7 programmable controllers provide organization block OB82 to detect and evaluate diagnostic interrupts. This OB is triggered when a DP slave with diagnostic capability detects an error (also called "event"). The CPU operating system calls OB82 in two situations – when a diagnostic interrupt has "come" and when it has "gone". However, two conditions must be met for this to be true: the DP slave must be capable of supporting this function, and you must have defined the diagnostic alarm in the parameter set for the DP slave using the *HW Config* program.

If you have not programmed OB82, the CPU reacts to a diagnostic interrupt by going into the STOP state. OB82 provides you with detailed information about the error that caused the interrupt on the DP slave. The 20 bytes of local data of OB82 (table 5.10) contain, among other things, the logical base address of the faulty DP slave or the faulty module of the DP slave, and four bytes of diagnostic information.

Table 5.10   Local data of OB82

Variable	Data Type	Description
OB82_EV_CLASS	BYTE	Interrupt class and identifiers: B#16#38 = going interrupt B#16#39 = coming interrupt
OB82_FLT_ID	BYTE	B#16#42 = error code
OB82_PRIORITY	BYTE	Priority class "26" (default value for RUN operational state) or "28" (STARTUP operational state)
OB82_OB_NUMBR	BYTE	OB number (82)
OB82_RESERVED_1	BYTE	Reserved
OB82_IO_FLAG	BYTE	B#16#54 = input module B#16#55 = output module
OB82_MDL_ADDR	INT	Logical base address of the module on which the error occurred
OB82_MDL_DEFECT	BOOL	Module fault
OB82_INT_FAULT	BOOL	Internal error
OB82_EXT_FAULT	BOOL	External error
OB82_PNT_INFO	BOOL	Channel error
OB82_EXT_VOLTAGE	BOOL	External auxiliary voltage missing
OB82_FLD_CONNCTR	BOOL	Front plug connector missing
OB82_NO_CONFIG	BOOL	Parameter set for module missing
OB82_CONFIG_ERR	BOOL	Wrong parameters on the module

Continued on page 95

Table 5.10 Continued

Variable	Data Type	Description
OB82_MDL_TYPE	BYTE	Bits 0 to 3: Module class Bit 4: Channel information present Bit 5: User information present Bit 6: Diagnostic interrupt from substitute Bit 7: In reserve
OB82_SUB_MDL_ERR	BOOL	User module wrong/missing
OB82_COMM_FAULT	BOOL	Communication error
OB82_MDL_STOP	BOOL	Operating state (0: RUN, 1: STOP)
OB82_WTCH_DOG_FLT	BOOL	Time monitoring has been triggered
OB82_INT_PS_FLT	BOOL	Internal module supply voltage has failed
OB82_PRIM_BATT_FLT	BOOL	Battery dead
OB82_BCKUP_BATT_FLT	BOOL	Entire battery backup failed
OB82_RESERVED_2	BOOL	Reserved
OB82_RACK_FLT	BOOL	Expansion rack failure
OB82_PROC_FLT	BOOL	Processor failure
OB82_EPROM_FLT	BOOL	EPROM error
OB82_RAM_FLT	BOOL	RAM error
OB82_ADU_FLT	BOOL	ADC/DAC error
OB82_FUSE_FLT	BOOL	Fuse blown
OB82_HW_INTR_FLT	BOOL	Process interrupt lost
OB82_RESERVED_3	BOOL	Reserved
OB82_DATE_TIME	DT	Date and time at which the OB was requested

5.2.7 Insert/Remove Module Interrupts (OB83)

The CPUs of the SIMATIC S7-400 series regularly check the presence of modules in the central rack and expansion rack. Modules that are plugged into DPV1 and S7 SP slaves, such as the ET 200M/IM 153-2 module, and are thus decentrally connected to the S7 system, are also verified by this monitoring function. If a configured module is removed from a modular DP slave while the S7 CPU is in the RUN state, the interrupt OB83 is triggered and, in addition, an entry is made in the diagnostic buffer and the module status list. If the module is removed from the slave while the S7 CPU is in the STOP state or is just starting up, the interrupt is registered in the diagnostic buffer and the module status list of the CPU, but OB83 is not called.

If a configured module is inserted while the CPU is running, the CPU checks whether the type of the inserted module matches the configuration. It then calls up OB83 and, provided that the module type is correct, the module is loaded with the parameter set which you have configured on the CPU using the *HW Config* program. At this point you may also call system functions (SFCs) to change the parameters of the reinserted module. Table 5.11 describes the local data of OB83.

Table 5.11 Local data of OB83

Variable	Data Type	Description
OB83_EV_CLASS	BYTE	Interrupt class and identifiers: B#16#38 = Module inserted B#16#39 = Module removed or cannot be addressed
OB83_FLT_ID	BYTE	Error code: (possible values: B#16#61, B#16#63, B#16#64, B#16#65)
OB83_PRIORITY	BYTE	Priority class "26" (default value for RUN state) or "28" (STARTUP state)
OB83_OB_NUMBR	BYTE	OB number (83)
OB83_RESERVED_1	BYTE	Reserved
OB83_MDL_ID	BYTE	B#16#54 = I/O area of the inputs (PE) B#16#55 = I/O area of the outputs (PA)
OB83_MDL_ADDR	WORD	Logical base address of the affected module
OB83_RACK_NUM	WORD	Number of the module rack or number of the DP station and DP master system ID (high byte)
OB83_MDL_TYPE	WORD	Module type of the affected module
OB83_DATE_TIME	DT	Date and time at which the OB was requested

A mismatch between the module type configured and the module type actually inserted in the DP slave is registered by local variable OB83_MDL_TYPE. Depending on the error code written to this variable, one of the following error messages is issued in this situation:

Table 5.12 Error codes reported by local variable OB83_FTL_ID

Error Codes in OB83_FLT_ID	Error, depending on the contents of OB83_MDL_TYPE
B#16#61	Module inserted, module type okay (for interrupt class B#16#38) Module removed or cannot be addressed (for interrupt class B#16#39) for OB83_MDL_TYPE = actual module type
B#16#63	Module inserted but wrong module type for OB83_MDL_TYPE = actual module type
B#16#64	Module inserted but faulty (type ID cannot be read) For OB83_MDL_TYPE = set module type
B#16#65	Module inserted but error in the module parameters For OB83_MDL_TYPE = actual module type

5.2.8 Program Sequence Error (OB85)

The operating system of the S7 CPU calls OB85 when the user program calls a block which is not loaded, or when the operating system calls an OB which was not programmed. OB85 is also called when an I/O access error occurs while the process image table is being updated. Such a situation can occur when the addresses configured for the inputs and outputs of a DP slave are located in the process image table of the S7 CPU and the DP slave has broken down. If you do not program OB85, the S7 CPU reacts to these errors by changing to the STOP state.

Table 5.13 shows the original contents of the local data of OB85. Table 5.14 is a suggestion of how to structure the local data of OB85 so that the user program can evaluate the error codes. The hexadecimal error codes "B1" and "B2" of the OB85_FTL_ID variables are particularly important when you use distributed I/O in your S7 system.

Table 5.13   Local data of OB85 (original structure)

Variable	Data Type	Description
OB85_EV_CLASS	BYTE	Interrupt class and identifiers (e.g., B#16#39 for "error while updating process image")
OB85_FLT_ID	BYTE	Error code: (Possible values: B#16#A1, B#16#A2, B#16#A3, B#16#B1, B#16#B2)
OB85_PRIORITY	BYTE	Priority class "26" (default value for RUN state) or "28" (STARTUP state)
OB85_OB_NUMBR	BYTE	OB number (85)
OB85_RESERVED_1	BYTE	Reserved
OB85_RESERVED_2	BYTE	Reserved
OB85_RESERVED_3	INT	Reserved
OB85_ERR_EV_CLASS	BYTE	Class of the interrupt which caused the error
OB85_ERR_EV_NUM	BYTE	Number of the interrupt which caused the error
OB85_OB_PRIOR	BYTE	Priority class of the OB which was being processed when the error occurred
OB85_OB_NUM	BYTE	Number of the OB which was being processed when the error occurred
OB85_DATE_TIME	DT	Date and time at which the OB was requested

Table 5.14 Structure of the local data of OB85 for error-code-dependent programming

Variable	Data Type
OB85_EV_CLASS	BYTE
OB85_FLT_ID	BYTE
OB85_PRIORITY	BYTE
OB85_OB_NUMBR	BYTE
OB85_DKZ23	BYTE
OB85_RESERVED_2	BYTE
OB85_Z1	WORD
OB85_Z23	DWORD
OB85_DATE_TIME	DATE_AND_TIME

Table 5.15 explains the error code reported by variable OB85_FLT_ID. The meaning of this error code depends on the contents of variables OB85_DKZ23, OB85_Z1 and OB85_Z23.

Table 5.15 OB85_FLT_ID error codes

Error Code OB85_FLT_ID	Meaning of Error
B#16#A1	In accordance with the STEP 7 program, the program or operating system generates a start event for an OB, but this OB is not loaded on the CPU.
B#16#A2	In accordance with the STEP 7 program, the program or operating system generates a start event for an OB, but this OB is not loaded on the CPU.
	The following additional information is provided by variables OB85_Z1 and OB85_Z23.
	OB85_Z1: Class of the event which caused the error (value of the interrupted program level)
	OB85_Z23: High word: Reports class and number of the triggering event Low word: Low word: Reports active program level and active OB at time of error
B#16#A3	Error while the operating system was accessing a block.
	The following additional information is provided by variables OB85_Z1 and OB85_Z23.
	OB85_Z1: Detailed error identifier of the operating system High byte: 1: Integrated function 2: IEC timer Low byte: 0: No error resolution 1: Block not loaded 2: Area length error 3: Write protection error

Continued on page 99

Table 5.15 Continued

Error Code OB85_FLT_ID	Meaning of Error
	OB85_Z23: High word: Block number Low word: Relative address of the MC7 command causing the error. For block type, see local variable OB85_DKZ23. B#16#88 = OB B#16#8C = FC B#16#8E = FB B#16#8A = DB
B#16#B1	I/O access error while updating the process image input table
B#16#B2	I/O access error while transferring the process image output table to the output modules
	The following additional information is provided by variables OB85_Z1 and OB85_Z23.
	OB85_Z1: Reserved for internal use by CPU OB85_Z23: Number of the I/O byte which caused the I/O access error (PZF)

5.2.9 Rack Failure (OB86)

The operating system of the S7 CPU reacts to a failure (coming event) or recovery (going event) of an expansion rack, DP master system or a DP slave by calling organization block OB86. If you do not program OB86, the S7 CPU changes to the STOP state when this type of event occurs.

Table 5.16 shows the original structure of the local data of OB86. The structure shown in table 5.17 is a suggestion on how to organize the local data of OB86 so that you can run a simple, error-code-dependent evaluation by means of the user program. The hexadecimal error codes "C3", "C4", and "C7" of variables OB86_FTL_ID are of particular importance when you use distributed I/O in your S7 system.

Table 5.16 Local data of OB86

Variable	Data Type	Description
OB86_EV_CLASS	BYTE	Event class and identifiers: B#16#38 = Going event B#16#39 = Coming event
OB86_FLT_ID	BYTE	Error code: (Possible values: B#16#C1, B#16#C2, B#16#C3, B#16#C4, B#16#C5, B#16#C6, B#16#C7)

Continued on page 100

Table 5.16   Continued

Variable	Data Type	Description
OB86_PRIORITY	BYTE	"26" (default value for RUN state) or "28" (STARTUP operational state)
OB86_OB_NUMBR	BYTE	OB number (86)
OB86_RESERVED_1	BYTE	Reserved
OB86_RESERVED_2	BYTE	Reserved
OB86_MDL_ADDR	WORD	Depends on error code
OB86_RACKS_FLTD	ARRAY [0..31] OF BOOL	Depends on error code
OB86_DATE_TIME	DT	Date and time at which OB was requested

Table 5.17   Local data structure of OB86 for error-code-dependent programming

Variable	Data Type
OB86_EV_CLASS	BYTE
OB86_FLT_ID	BYTE
OB86_PRIORITY	BYTE
OB86_OB_NUMBR	BYTE
OB86_RESERVED_1	BYTE
OB86_RESERVED_2	BYTE
OB86_MDL_ADDR	WORD
OB86_Z23	DWORD
OB86_DATE_TIME	DATE_AND_TIME

Table 5.18 explains the error code reported by variable OB86_FLT_ID. The meaning of this error code depends on the contents of variables OB86_DKZ23, OB86_Z1 and OB86_Z23.

Table 5.18 OB86_FLT_ID error codes

Error Code OB86_FLT_ID	Meaning of Error

B#16#C1	Expansion rack failure
	OB86_MDL_ADDR: Logical base address of the IM
	
	The following additional information is provided by variable OB86_Z23.
	
	OB86_Z23: One bit is assigned to every possible expansion rack.
	      Bit 0: Always 0
	      Bit 1: 1st expansion rack
	      : :
	      : :
	      Bit 21: 21st expansion rack
	      Bit 22 to 29: Always 0
	      Bit 30: Failure of at least one expansion rack in the SIMATIC S5 area
	      Bit 31: Always 0
	
	Remarks:
	
	The coming event indicates the failure of the expansion racks (the assigned bits are set to "1"). It always refers to those expansion racks that triggered the call of OB86. Expansion racks which already failed earlier are no longer indicated.
	The going event indicates the recovery of expansion racks failed earlier on (the assigned bits are set to "1").

B#16#C2	Recovery of an expansion rack with identifier:
	"Failure of an expansion rack, *going*, with deviation in set/actual configuration"
	
	OB86_MDL_ADDR: Logical base address of the IM
	
	The following additional information is provided by variable OB86_Z23.
	
	OB86_Z23: Contains one bit for every possible expansion rack (see error code B#16#C1). Meaning of a set bit:
	      On the affected expansion rack:
	      – Modules with wrong type identifier present
	      – Configured modules missing
	      – At least one module defective

B#16#C3	Failure of a DP master system for distributed I/O. (A coming event supplies error code B#16#C3; a going event supplies error code B#16#C4 and event class B#16#38). The recovery of each lower-level DP station also starts OB86.
	
	OB86_MDL_ADDR: Logical base address of the DP master
	
	The following additional information is provided by variable OB86_Z23.
	
	OB86_Z23: DP master system ID
	      Bit 0 to 7: Reserved
	      Bit 8 to 15: DP master system ID
	      Bit 16 to 31: Reserved

Continued on page 102

Table 5.18 Continued

Error Code OB86_FLT_ID	Meaning of Error
B#16#C4	Failure of a DP station
B#16#C5	Malfunction of a DP station
	OB86_MDL_ADDR: Logical base address of the DP master
	The following additional information is provided by variable OB86_Z23.

	OB86_Z23: Address of the affected DP slave
	          Bit 0 to 7 : Number of the DP station
	          Bit 8 to 15: DP master system ID
	          Bit 16 to 30: Logical base address for an
	                                       S7 DP slave or diagnostic
	                                       address for a standard DP slave
	          Bit 31: I/O identifier

B#16#C6	Recovery of an expansion rack, but error in module parameter set
	OB86_MDL_ADDR: Logical base address of the IM
	The following additional information is provided by variable OB86_Z23.

	OB86_Z23: One bit is assigned to every possible expansion rack.
	          Bit 0: Always 0
	          Bit 1: 1st expansion rack
	          : :
	          : :
	          Bit 21: 21st expansion rack
	          Bit 22 to 30: Reserved
	          Bit 31: Always 0

	Meaning of a set bit:

	The affected expansion rack has modules with
	   – wrong type identifier or
	   – missing or wrong parameters

B#16#C7	Recovery of a DP station but error in module parameter set
	OB86_MDL_ADDR: Logical base address of the DP master
	The following additional information is provided by variable OB86_Z23.

	OB86_Z23: Address of the affected DP slave:
	          Bit 0 to 7: Number of the DP station
	          Bit 8 to 15: DP master system ID
	          Bit 16 to 30: Logical base address
	                                       of the DP slave
	          Bit 31: I/O identifier

5.2.10 I/O Access Error (OB122)

The operating system of the S7 CPU calls OB122 when an error occurs while a STEP 7 instruction is attempting to gain access to the input/output data of an I/O module or a DP slave. OB122 is also called when the user program attempts to gain access to an input or output of a non-existent or defective DP slave. If you do not program OB122, the CPU reacts to such an I/O access error by switching to the STOP state. Table 5.19 shows the local data of OB122.

Table 5.19 Local data of OB122

Variable	Data Type	Description
OB122_EV_CLASS	BYTE	Event class and identifiers (e.g., B#16#29 for "I/O access error")

OB122_SW_FLT	BYTE	Error code
		B#16#42 = (For S7-300) error during read access to I/O
		       = (For S7-400) error during first read access to I/O after occurrence of an error
		B#16#43 = (S7-300) error during write access to I/O
		       = (S7-400) error during first write access to I/O after occurrence of an error
		B#16#44 = (Only for S7-400) error during nth (n > 1) read access to I/O after occurrence of an error
		B#16#45 = (Only for S7-400) error during nth (n > 1) write access to I/O after occurrence of an error

OB122_PRIORITY	BYTE	Priority class of the OB in which the error occurred
OB122_OB_NUMBR	BYTE	OB number (122)
OB122_BLK_TYPE	BYTE	Block type in which the error occurred B#16#88 = OB B#16#8A = DB B#16#8C = FC B#16#8E = FB

OB122_MEM_AREA	BYTE	Access type and memory area
		Bits 7 to 4, access type:
		  0: Bit access
		  1: Byte access
		  2: Word access
		  3: Double word access
		Bits 3 to 0, memory area:
		  0: I/O area
		  1: Process image input table
		  2: Process image output table

OB122_MEM_ADDR	WORD	Memory address at which the error occurred
OB122_BLK_NUM	WORD	Number of the block with the MC7 command which is causing the error
OB122_PRG_ADDR	WORD	Relative address of the MC7 command which is causing the error
OB122_DATE_TIME	DT	Date and time at which OB was requested

5.3 DP User Data Communication and Process Interrupt Functions

5.3.1 Exchange of Consistent DP Data with SFC14 *DPRD_DAT* and SFC15 *DPWR_DAT*

The I/O access commands available in STEP 7 do not permit access to DP data areas (modules) which have a closed (consistent) structure of 3 or >4 bytes using simple byte, word or double-word commands. See also section 6.1. To access such coherent data areas, use system functions *DPRD_DAT* and *DPWR_DAT*.

System function SFC 14 *DPRD_DAT*

To read a consistent input data area (see section 6.2) of a DP slave, use the system function SFC14 *DPRD_DAT*. Each read access refers to a specific input module. If a DP slave has several consistent input modules, you must program a separate SFC14 call for every input module you want to read. Table 5.20 lists the input and output parameters of SFC14 which you must define.

Table 5.20   Parameters for SFC14 *DPRD_DAT*

Parameter	Declaration	Data Type	Memory Area	Description
LADDR	INPUT	WORD	I, Q, M, D, L, constant	Specification (in hex format) of the start address of the input modules of the DP slave configured with *HW Config*
RET_VAL	OUTPUT	INT	I, Q, M, D, L	Return value of the SFC
RECORD	OUTPUT	ANY	I, Q, M, D, L	Destination area for the user data read

Parameter description

RECORD

The RECORD parameter describes the destination area on the S7 CPU for the consistent input data read from the DP slave. The length you define here must match the length which you have defined in the *HW Config* program for the input module on the DP slave. Please also note that the RECORD parameter is of data type ANY-Pointer. For ANY-Pointer only the data type BYTE is allowed.

RET_VAL

The error codes of the RET_VAL parameter of system function SFC14 are shown in table 5.21.

Table 5.21 Return values of the RET_VAL parameter of SFC14 *DPRD_DAT*

Error Code w#16#...	Explanation
0000	No error occurred.
8090	No module is configured for the specified logical base address or the permitted length for consistent data was exceeded or the initial address at the LADDR parameter was entered in hexadecimal format.
8092	A type other than BYTE specified in parameter of data type ANY-Pointer
8093	No DP module from which to read consistent data exists for the logical address specified under LADDR.
80A0	An access error is detected when the module is accessed
80B0	Slave failure on external DP interface
80B1	The length of the specified destination area does not match the length of user data specified in *HW Config*.
80B2 80B3 80C0 80C2 80Fx 87xy 808x	System error for external DP interface System error for external DP interface System error for external DP interface System error for external DP interface System error for external DP interface System error for external DP interface System error for external DP interface

System function SFC15 *DPWR_DAT*

To transfer a consistent output data area from the S7 CPU to the DP slave, use the system function SFC15 *DPWR_DAT*. Each such write access refers to a specific output module. If a DP slave has several consistent output modules, you must program a separate SFC15 call for every output module you want to write to. Table 5.22 shows the input and output parameters of SFC15 which you must define.

Table 5.22 Parameters for SFC15 *DPWR_DAT*

Parameter	Declaration	Data Type	Memory Area	Description
LADDR	INPUT	WORD	I, Q, M, D, L, constant	Specification (in hex format) of the start address of the output module of the DP slave configured in *HW Config*
RECORD	OUTPUT	ANY	I, Q, M, D, L	Source area of the user data to be written
RET_VAL	OUTPUT	INT	I, Q, M, D, L	Return value of the SFC

Parameter description

RECORD

The RECORD parameter describes the source area of the consistent output data to be written from the S7 CPU to the DP slave. The length which you specify here must match the length of the output module of the DP slave configured in *HW Config*. Please also note that the RECORD parameter is of data type ANY-Pointer. For ANY-Pointer only the data type BYTE is allowed.

RET_VAL

The error codes of the RET_VAL parameter of SFC15 are listed in table 5.23.

Table 5.23   Specific return values for SFC15 *DPWR_DAT*

Error Code w#16#...	Explanation
0000	No error has occurred.
8090	No module is configured for the specified logical base address or the permitted length of consistent data was exceeded.
8092	A type other than BYTE is specified in the parameter of data type ANY-Pointer.
8093	No DP module to which consistent data can be written exists for the logical address specified under LADDR.
80A1	The selected module is faulty.
80B0	Slave failure on an external DP interface
80B1	The length of the specified source area does not match the length specified for the user data configured in *HW Config*.
80B2	System error for external DP interface
80B3	System error for external DP interface
80C1	The data of the preceding write job on the module has not yet been processed by the module.
80C2 80Fx 85xy	System error for external DP interface System error for external DP interface System error for external DP interface

5.3.2 Control Commands SYNC and FREEZE Transmitted with SFC11 *DPSYC_FR*

Use system function SFC11 *DPSYC_FR* if you want to transfer the control commands SYNC and FREEZE to one or more DP slaves. These commands synchronize data communication with certain DP slaves. The DP slaves concerned must be combined in SYNC/FREEZE groups during configuration.

A global control telegram (Broadcast telegram) sends the SYNC and FREEZE control commands simultaneously to all DP slaves.

DP control command SYNC

The SYNC control command synchronizes the outputs of DP slaves. When a DP slave is in SYNC mode, the output data transferred with the Data_Exchange telegram is stored in a local transfer buffer of the DP slave. After receiving the SYNC control command, the DP slave transfers the data stored in the transfer buffer to the outputs. In this way, the SYNC command ensures simultaneous activation (synchronization) of output data of several DP slaves.

DP control command UNSYNC

The UNSYNC control command cancels the SYNC mode of the addressed DP slaves so that they can again participate in cyclic data transfer with the DP master. The output data received with the Data_Exchange telegram is thus forwarded immediately to the outputs on the DP slave.

DP control command FREEZE

The FREEZE control command freezes the inputs of DP slaves. When a DP slave is operated in FREEZE mode and it receives a FREEZE command from the DP master, the currently queued input data of the DP slave is stored in a transfer memory on the DP slave and thus frozen. The DP master then can send a Data_Exchange telegram to read this frozen input data from the transfer memory of the DP slave. The data currently queued on inputs of the DP slave is only read when another FREEZE command is received. Again, this data is frozen in the transfer memory of the DP slave. And again, the DP master can keep on reading this frozen data until the next FREEZE command is received. This control command permits simultaneous (synchronous) transmission of the input data currently queued on the DP slaves.

DP control command UNFREEZE

The UNFREEZE control command cancels the FREEZE mode of the addressed DP slaves so that they can again participate in cyclic data transfer with the DP master. The input data of a DP slave is no longer intermediately stored in a buffer. It can now be read immediately by the DP master.

Table 5.24 lists the SFC11's input and output parameters which must be supplied during the SFC call.

Parameter description

GROUP

During the configuration phase with the *HW Config* program you have already assigned the DP slaves to a certain group. With the GROUP parameter you now specify which group will be addressed with SFC11. You can activate several groups with one job. The value "0" (all bits set to "0") cannot be used as a group. Table 5.25 describes the group assignment.

Table 5.24 Parameters of SFC11 *DPSYC_FR*

Parameter	Declaration	Data Type	Memory Area	Description
REQ	Input	BOOL	I, Q, M, D, L, constant	REQ = "1" triggers a SYNC/FREEZE job
LADDR	Input	WORD	I, Q, M, D, L, constant	Logical base address of the DP master
GROUP	Input	BYTE	I, Q, M, D, L, constant	GROUP selector Bit x = 0: Group not affected Bit x = 1: Group affected
MODE	Input	BYTE	I, Q, M, D, L, constant	Job identifier
RET_VAL	Output	INT	I, Q, M, D, L	Return value of the SFC
BUSY	Output	BOOL	I, Q, M, D, L	BUSY = "1" means that the triggered SFC11 *DPSYC_FR* has not yet been concluded.

Table 5.25 Group assignment of SFC11 *DPSYC_FR* in GROUP parameter

	Bit 7	Bit 6	Bit 5	Bit 4	Bit 3	Bit 2	Bit 1	Bit 0
GROUP	8	7	6	5	4	3	2	1

MODE

Use the MODE parameter to assign and transfer a control command to a group. Table 5.26 shows which control bits of the MODE parameter set which mode.

Table 5.26 Control commands in the MODE parameter of SFC11 *DPSYC_FR*

Bit No:	7	6	5	4	3	2	1	0
MODE			SYNC	UNSYNC	FREEZE	UNFREEZE		

With each SFC11 call, you can set more than one control command and send them to the DP slaves. Table 5.27 shows you the possible combinations. The MODE parameter allows you to sent several control commands to a DP slave with just one SFC11 call.

RET_VAL

Table 5.28 lists and describes the error codes of the RET_VAL parameter with SFC11.

Table 5.27 Possible combinations of the MODE parameter of SFC11 *DPSYC_FR*

Bit No:	7	6	5	4	3	2	1	0
				UNSYNC				
				UNSYNC		UNFREEZE		
				UNSYNC	FREEZE			
MODE			SYNC					
			SYNC			UNFREEZE		
			SYNC		FREEZE			
						UNFREEZE		
					FREEZE			

Table 5.28 Error codes of SFC11 *DPSYC_FR*, stored in parameter RET_VAL

Error Code W#16#...	Explanation
0000	The job was performed correctly.
7000	First call with REQ = "0". No SFC11 *DPSYC_FR* active; BUSY has the value "0".
7001	First call with REQ = "1". A process interrupt request was sent to the DP master; BUSY has the value "1".
7002	Intermediate call (REQ irrelevant): The triggered SFC11 DP_SYC_FR has not yet been concluded; BUSY has the value "1".
8090	Specified logical base address is invalid. LADDR is not a DP master.
8093	This SFC cannot be used for the module selected in LADDR.
80B0	Group not configured
80B1	Group not assigned to this CPU
80B2	SYNC mode not permitted for this group
80B3	FREEZE mode not permitted for this group
80C2	At the moment, the module is processing the maximum number of jobs for one CPU. All resources for this CPU are occupied.
80C5	Distributed I/O not available DP subsystem failure (bus error or ET-CR operating mode switch is on STOP).
80C6	I/O rejected by CPU (job terminated)
80C7	Termination due to a warm restart of ET-CR. Hot restart not possible.
8325	GROUP parameter wrong
8425	MODE parameter wrong

5.3.3 Triggering a Process Interrupt on the DP Master with SFC7 *DP_PRAL*

A SIMATIC S7-300 programmable controller which acts as an I-slave on the DP bus and uses a CPU31x-2DP for this purpose, is able to trigger a process interrupt on the DP master system. System function SFC7 *DP_PRAL* is used for this purpose.

A CPU31x-2DP configured as an I-slave can call system function SFC7 *DP_PRAL* and thus trigger a process interrupt (OB40 to OB47) on the related DP master (only S7-400 and S7-300 with CPU31x-2DP) from within the user program. The SFC input parameter AL_INFO can be used to transfer a user-related interrupt identifier. This interrupt identifier is transferred to the DP master (variable *OB40_POINT_ADDR*). There it can be evaluated by an interrupt OB (OB40 to OB47). The parameters IOID and LADDR unambiguously define the requested process interrupt. Exactly one process interrupt can be triggered at any time for each configured module in the transfer memory of the I-slave. Table 5.29 shows the input and output parameters of SFC7.

Table 5.29 Parameters of SFC7 *DP_PRAL*

Parameter	Declaration	Data Type	Memory Area	Description
REQ	INPUT	BOOL	I, Q, M, D, L, constant	Request to trigger a process interrupt on DP master.
IOID	INPUT	WORD	I, Q, M, D, L, constant	Identifier of the address area (module) in the transfer memory (from viewpoint of the DP slave) B#16#54 = I/O input (PE) B#16#55 = I/O output (PA) The identifier of an area which belongs to a composite module is the identifier of the lower of the two addresses. If addresses are the same, specify B#16#54.
LADDR	INPUT	WORD	I, Q, M, D, L, constant	Start address of the address area (module) in the transfer memory (from viewpoint of the DP slave). If this is an area which belongs to a composite module, specify the lower of the two addresses.
AL_INFO	INPUT	DWORD	I, Q, M, D, L, constant	Interrupt identifier This is supplied to OB40 which is to be started on the applicable DP master (variable OB40_ POINT_ADDR).

Continued on page 111

Table 5.29 Continued

Parameter	Declaration	Data Type	Memory Area	Description
RET_VAL	OUTPUT	INT	I, Q, M, D, L	Return value of the SFC
BUSY	OUTPUT	BOOL	I, Q, M, D, L	BUSY = "1" means that the triggered SFC7 *DP_PRAL* has not yet been acknowledged by the DP master.

SFC7 *DP_PRAL* is processed asynchronously (i.e., processing takes place over several SFC calls and therefore also several CPU cycles). The job is finished when the process interrupt is acknowledged by the DP master after the appropriate interrupt OB (OB40 to OB47) has been completely processed.

If the CPU31x-2DP is a standard DP slave, the SFC7 job is concluded as soon as the diagnostic telegram has been fetched by the DP master. Table 5.30 shows possible error codes of SFC7 which are indicated by the RET_VAL parameter.

Table 5.30 Specific return values for SFC7 *DP_PRAL*

Error Code W#16#...	Explanation
0000	The job was performed correctly.
7000	First call with REQ = "0". No process interrupt request is active. BUSY has the value "0".
7001	First call with REQ = "1". A process interrupt request was sent to the DP master. BUSY has the value "1".
7002	Intermediate call (REQ irrelevant): The triggered process interrupt has not yet been acknowledged by the DP master. BUSY has the value "1".
8090	Start address of the address area in the transfer memory is wrong.
8091	Interrupt disabled by the configuration
8093	The parameter pair IOID and LADDR is used to address a module from which no process interrupt request can be made.
80C6	Distributed I/O is not available at the moment.

5.4 Reading DP Diagnostic Data

5.4.1 Reading the Standard Diagnostic Data of a DP Slave with SFC13 *DPNRM_DG*

DP slaves provide diagnostic data for detection and localization of local errors. The principal layout of the diagnostic data of a DP slave is specified in volume 2 of the EN 50 170 standard (PROFIBUS), and is listed in table 5.31. For additional detailed information on the diagnostic data, see also section 7 on the diagnostic functions.

Table 5.31 Principal layout of the DP slave diagnostic data

Byte 0	Station status 1
Byte 1	Station status 2
Byte 2	Station status 3
Byte 3	PROFIBUS address of the DP master
Byte 4	Manufacturer's identifier (high byte)
Byte 5	Manufacturer's identifier (low byte)
ab Byte 6	Additional slave-related diagnostic data

The diagnostic data of a DP slave can be read by means of SFC13 *DPNRM_DG*. Table 5.32 specifies the input and output parameters of SFC13.

Table 5.32 Parameters of SFC13 *DPNRM_DG*

Parameter	Declaration	Data Type	Memory Area	Description
REQ	INPUT	BOOL	I, Q, M, D, L, constant	Request to read
LADDR	INPUT	WORD	I, Q, M, D, L, constant	The diagnostic address of the DP slave configured in *HW Config* (specification in hexadecimal format)
RET_VAL	OUTPUT	INT	I, Q, M, D, L	Return value of the SFC (error message or length in bytes of the diagnostic data read)
RECORD	OUTPUT	ANY	I, Q, M, D, L	Destination area for the diagnostic data read
BUSY	OUTPUT	BOOL	I, Q, M, D, L	BUSY = "1": Read procedure concluded

SFC13 is processed asynchronously. This means that execution of the function may take place over several SFC calls and therefore also several CPU cycles.

The SFC13-related error codes reported with the RET_VAL parameter are a subset of the error codes for SFC59. See section 5.9.

Parameter description

RECORD

The RECORD parameter describes the destination area on the CPU for the diagnostic data read from the DP slave. It is a type ANY-Pointer parameter and only permits the data type BYTE.

If the number of bytes of diagnostic data which were read from a DP slave is greater than the specified destination area, the diagnostic data are rejected and an appropriate error code is transmitted by the RET_VAL parameter. If the length of the diagnostic data read is equal to or less than the length specified for the RECORD parameter, the data which was read is accepted in the destination area and the actual number of bytes read is reported with the RET_VAL parameter.

The minimum length of the diagnostic data to be read is 6 bytes. The maximum length is 240 bytes. If a DP slave supplies more than 240 bytes of diagnostic data (up to 244 bytes are permitted) and a destination area of this length is reserved by the RECORD parameter, the first 240 bytes are transferred to the destination area and the overflow bit is set in the diagnostic data. If a DP slave supplies more than 240 bytes of diagnostic data and the length specified by the RECORD parameter is less than 240 bytes, the entire diagnostic telegram is rejected.

System resources for SFC13 in SIMATIC S7-400 systems

When, in an S7-400 system, system function SFC13 *DPNRM_DG* is called repeatedly, resources (memory space) are allocated for the asynchronous job processing of the SFC. When several jobs are active at the same time, it is ensured that all jobs are performed and do not affect each other. However, only a certain number of SFC13 calls may be active simultaneously. For the maximum number of possible SFC calls, see the performance data of the S7-400 CPU you are using. If the maximum amount of usable resources is reached, an error code is transmitted by the RET_VAL parameter. If this happens, the SFC must be triggered again.

5.4.2 Interrupt from a DP Slave Received with SFB54 *RALRM*

SFB54 *RALRM* receives an interrupt, including the associated information, from a module of a DP slave.

The output parameters of SFB54 provide information on the interrupt source, the start information of the called OB and the interrupt information of the interrupt-generating DP slave.

SFB54 may only be called within the interrupt OB started by the operating system of the CPU as a result of the trigger event from the distributed I/O.

The interface of SFB54 *RALRM* is identical with that of the FB *RALRM* defined in the PNO AK 1131 standard. The input and output parameters of SFB54 are shown in Table 5.33.

Table 5.33 Parameters of SFB54 *RALRM*

Parameter	Declaration	Data type	Memory area	Description
MODE	INPUT	INT	I, Q, M, D, L, constant	Operating mode of SFB54
F_ID	INPUT	DWORD	I, Q, M, D, L, constant	Specification (in hexadecimal format) of the starting address, configured in *HW Config*, of the module of the DP slave from which interrupts are expected
MLEN	INPUT	INT	I, Q, M, D, L, constant	Maximum length in bytes of the interrupt information to be received
NEW	OUTPUT	BOOL	I, Q, M, D, L	TRUE = a new interrupt has been received
STATUS	OUTPUT	DWORD	I, Q, M, D, L	Error code of the SFB or the DP master
ID	OUTPUT	DWORD	I, Q, M, D, L	Logical starting address of the interrupt-generating module
LEN	OUTPUT	INT	I, Q, M, D, L	Length of the received interrupt information
TINFO	OUTPUT	ANY	I, Q, M, D, L	*Task information* OB start information and administrative information
AINFO	OUTPUT	ANY	I, Q, M, D, L	*Interrupt information* Header information and additional interrupt information

Description of parameters

MODE

The SFB54 *RALRM* can be called in different operating modes (*MODE*). Table 5.34 lists and describes the different modes.

Table 5.34 Structure of the administrative information

MODE	Meaning
0	SFB54 indicates the interrupt-generating module in output parameter ID and overwrites the output parameter NEW with TRUE
1	SFB54 writes all output parameters independently of the interrupt-generating module
2	SFB54 checks to see if the component specified in the input parameter F_ID generated the interrupt If yes: The parameter NEW is overwritten with TRUE and the relevant values are written to all other output parameters If no: The parameter NEW is overwritten with FALSE

TINFO

The TINFO parameter specifies the destination area for the OB start information and the administrative information. If the selected destination area is too small, SFB54 cannot enter all the information.

The OB start information, in which SFB54 is called, is entered from byte 0 to byte 19, and the administrative information is entered from byte 20 to byte 27 (see Table 5.35).

Table 5.35 Structure of the administrative information

Byte no. for TINFO	Data type	Meaning	
20	BYTE	DP master system ID (possible values 1 to 255)	
21	BYTE	Address of the DP slave	
22	BYTE	Bit 0 to 3: slave type	0000 = DP slave 0001 = S7 DP slave 0010 = DPS7V1 slave 0011 = DPV1 slave from 0100 = reserved
		Bit 4 to 7: profile type	0000 = DP from 0001 = reserved
23	BYTE	Bit 0 to 3: interr. info type	0000 = transparent (interrupt from a configured distributed module) 0001 = proxy (interrupt not from a DPV1 slave or configured slot) 0010 = Interrupt generated in the CPU from 0011 = reserved
		Bit 4 to 7: Structure version	0000 = initial value from 0001 = reserved

24	BYTE	Flags of the DP master interface
		    Bit 0 = 0 → interrupt from an integrated DP master
		    Bit 0 = 1 → interrupt from an external DP master
		    Bit 1 to 7 = reserved	

25	BYTE	Flags of the DP slave interface
		    Bit 0: EXT_DIAG_BIT from the diagnostisc frame
		        or 0, if this bit not available at time of interrupt
		    Bit 1 to 7 = reserved	

26 to 27	WORD	PROFIBUS identification number of the slave	

AINFO

The AINFO parameter defines the destination area for the header information and the additional interrupt information. If the selected destination area is too small, SFB54 cannot enter all the information. For this reason, a length of at least MLEN bytes must be parameterized for AINFO.

The header information is entered from byte 0 to Byte 3 (see Table 5.36), and the additional interrupt information is entered from byte 4 to no higher than byte 63 (in the case of interrupts from distributed I/O).

Table 5.36 Structure of the header information

Byte no. for AINFO	Data type	Meaning
0	BYTE	Length of the received interrupt information in bytes (4 to 63)
1	BYTE	ID for the interrupt type 1: Diagnostic interrupt 2: Process interrupt 3: Remove interrupt 4: Insert interrupt 5: Status interrupt 6: Update interrupt 31: Failure of an expansion unit, a DP master system, or a DP station 32 to 126: Vendor-specific interrupt
2	BYTE	Slot number of the interrupt-generating component
3	BYTE	Specifier 0: No other information 1: Coming event, slot fault 2: Going event, slot fault eliminated 3: Going event, slot fault continues

Destination area TINFO and AINFO

Depending on the OB in which SFB54 is called, the destination areas TINFO and AINFO are only partly overwritten. Table 5.37 lists the information that is entered in each case.

Table 5.37 Availability of the interrupt information

Interrupt type	OB	TINFO OB status information	TINFO Admin. information	AINFO Header information	AINFO Additional interrupt information
Process interrupt	4x	Yes	Yes	Yes	As supplied by DP slave
Status interrupt	55	Yes	Yes	Yes	Yes
Update interrupt	56	Yes	Yes	Yes	Yes
Vendor-specific interrupt	57	Yes	Yes	Yes	Yes
Diagnostic interrupt	82	Yes	Yes	Yes	As supplied by DP slave
Insert/remove interrupt	83	Yes	Yes	Yes	As supplied by DP slave
Station failure	86	Yes	Yes	No	No

STATUS

The STATUS output parameter contains fault information. If it is interpreted as ARRAY[1...4] OF BYTE, the fault information has the structure shown in Table 5.38.

Table 5.38 Representation of the STATUS output parameter

Field element	Name	Meaning
STATUS[1]	Function_Num	B#16#00, if no error Function ID from DPV1-PDU: In the event of an error, B#16#80 is prompted. If no DPV1 protocol element is used: B#16#C0.
STATUS[2]	Error_Decode	Location of error code
STATUS[3]	Error_Code_1	Error code
STATUS[4]	Error_Code_2	Vendor-specific expansion of the error code

The location of the error code is entered in STATUS[2] and shown in Table 5.39. The error code itself is in STATUS[3] and is represented in Table 5.40.

The error code in STATUS[4] is passed from the DP master to the CPU in the case of DPV1 errors. If there is no DPV1 error, the value is set to "0".

Table 5.39 Error codes in STATUS[2]

Error_Decode (B#16#...)	Source	Meaning
00 to 7F	CPU	No error or no warning
80	DPV1	Error in accordance with IEC 61158-6
81 to 8F	CPU	B#16#8x shows an error in the nth call parameter of the SFB
FE, FF	DP Profile	Profile-specific error

Table 5.40 Error codes in STATUS[3]

Error_Decode (B#16#...)	Error_Code_1 (B#16#...)	Explanation according to DVP1	Meaning
00	00		No error, no warning
70	00	Reserved, reject	Initial call; no data set transfer active
	01	Reserved, reject	Initial call; data set transfer initiated
	02	Reserved, reject	Intermediate call; data set transfer already active
80	90	Reserved, pass	Logical starting address invalid
	92	Reserved, pass	Impermissible type at ANY pointer
	93	Reserved, pass	The DP component addressed using ID or F_ID is not configured
	A0	Read error	Negative acknowledgement when reading from the module
	A1	Write error	Negative acknowledgement when writing to module
	A2	Module failure	DP protocol error on Layer 2, possible hardware defect
	A3	Reserved, pass	DP protocol error on direct data link mapper or user-interface/ user, possible hardware defect
	A4	Reserved, pass	Communication fault on K bus
	A5	Reserved, pass	–
	A7	Reserved, pass	DP resource in use
	A8	Version conflict	Version conflict
	A9	Feature not supported	Feature not supported
	AA to AF	User specific	Specific to DP master
	B0	Invalid index	Module does not know the data set: data set number ≥ 256 not permissible

Continued on page 119

Table 5.40 Continued

Error_Decode (B#16#...)	Error_Code_1 (B#16#...)	Explanation according to DVP1	Meaning
	B1	Write length error	Length error in AINFO
	B2	Invalid slot	The configured slot is not occupied
	B3	Type conflict	Actual module type not identical to desired module type
	B4	Invalid area	Invalid area
	B5	State conflict	State conflict
	B6	Access denied	Access denied
	B7	Invalid range	Invalid range
	B8	Invalid parameter	Invalid parameter
	B9	Invalid type	Invalid type
	BA to BF	User specific	Specific to DP master
	C0	Read constrain conflict	Module has the data set but there are no read data yet
	C1	Write constrain conflict	Data of the previous write job on the module for the same data set have not yet been processed by the module
	C2	Resource busy	Module currently processing the possible maximum of jobs for a CPU
	C3	Resource unavailable	The required resource is currently in use
	Dx	User specific	Specific to DP slave, see description of DP slave
81	00 to FF		Error in first call parameter (in case of SFB54: MODE)
	00		Impermissible mode
82	00 to FF		Error in second call parameter
↓	↓		↓
88	00 to FF		Error in eighth call parameter (in case of SFB54: TINFO)
	01		Wrong syntax ID
	23		Quantity structure exceeded or destination area too small
	24		Wrong area ID
	32		DB/DI no. outside user range

Continued on page 120

Table 5.40 Continued

Error_Decode (B#16#...)	Error_Code_1 (B#16#...)	Explanation according to DVP1	Meaning
	3A		DB/DI no. is ZERO in the case of area ID DB/DI or specified DB/DI is not available
89	00 to FF		Error in ninth call parameter (in case of SFB54: AINFO)
	01		Wrong syntax ID
	23		Quantity structure exceeded or destination area too small
	24		Wrong area ID
	32		DB/DI no. outside user range
	3A		DB/DI no. is ZERO in the case of area ID DB/DI, or specified DB/DI is not available
8A	00 to FF		Error in 10th call parameter
↓	↓		↓
8F	00 to FF		Error in 15th call parameter
FE, FF	00 to FF		Profile-specific error

5.4.3 DP-Related System State List (SZL)

The system state list (SZL) is an information function. It describes the current status of the automation system. The system status list can only be read but not changed. The DP-related system status sublists are virtual lists. This means that they are only generated by the operating system on request.

System state lists contain the following information.

▷ *System data*
System data contains the fixed and the adjustable characteristic data of a CPU. System data describes the hardware configuration of the CPU, the state of the priority classes and the communication.

▷ *Diagnostic state data on the CPU*
Diagnostic state data describes the current status of all components which are monitored by the system diagnostic functions.

▷ *Diagnostic data on modules*
Modules that are assigned to a CPU generate and store diagnostic data provided that they have diagnostic capability.

▷ *Diagnostic buffer*
All diagnostic events are entered in the diagnostic buffer in the order in which they occurred.

5.4.4 Layout of a System Status Sublist

A system status sublist always consists of a header and the data records actually requested. The header of the sublist contains the ID of the system status sublist (SZL ID), the index, the length in bytes of a data record of the requested sublist and the number of data records which this sublist contains. The data record of a sublist has a certain length which is based on the information and structure.

5.4.5 Reading System Status Sublists with SFC51 *RDSYSST*

To read the contents of a system status sublist, or an excerpt of it, use system function SFC51 *RDSYSST* (*ReaD SYStem STatus*). The parameters SZL_ID and INDEX of SFC51 determine which sublist or which sublist excerpt will be read.
Table 5.41 lists the call parameters for SFC51 *RDSYSST*.

Table 5.41 Parameters of SFC51 *RDSYSST*

Parameter	Declaration	Datentyp	Data Type	Description
REQ	INPUT	BOOL	I, Q, M, D, L, constant	REQ = "1": Trigger for processing
SZL_ID	INPUT	WORD	I, Q, M, D, L, constant	SSL ID of the sublist or the sublist excerpt
INDEX	INPUT	WORD	I, Q, M, D, L, constant	Type or number of an object in a sublist
RET_VAL	OUTPUT	INT	I, Q, M, D, L	Return value of the SFC
BUSY	OUTPUT	BOOL	I, Q, M, D, L	BUSY = "1": Read procedure not yet concluded
SZL_HEADER	OUTPUT	STRUCT	D, L	See parameter SZL_HEADER
DR	OUTPUT	ANY	I, Q, M, D, L	Field of the read data records

Parameter description

SZL_ID

Each sublist of a system status list has its own identifying number (SZL_ID). Figure 5.3 shows the layout of this identifying number. Possible SZL_IDs are listed in table 5.44 in section 5.4.6. To request a complete system status sublist, or an excerpt of it, specify the relevant ID.

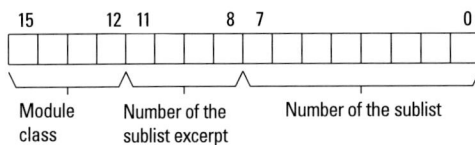

Figure 5.3 Layout of the SZL-ID

The possible excerpts of a system status sublist are predefined and cannot be changed. They, too, are identified by a number. The identifying number of the sublist excerpt and its meaning depend on the requested sublist.

The identifying number SZL_ID contains 4 additional bits which identify the module class. These bits specify the type of module from which the sublist or sublist excerpt is to be read. Examples of which module classes can be assigned to which IDs are given in table 5.42.

Table 5.42 Assignment of module classes to identifier numbers

Identifier Number/Bits	Module Type
0000	CPU
1100	CP
0100	IM
0101	Analog module
1111	Digital module

An ID of the system status list (SZL-ID) consists of the number of the sublist, the number of the sublist excerpt and the module class (see Figure 5.3).

INDEX

Certain sublists or sublist excerpts require an object type identifier or an object number. In this case you have to use the INDEX parameter. In situations where this parameter is not necessary you can disregard its contents.

RET_VAL parameter

The error codes transmitted by RET_VAL parameter are described in table 5.43.

Table 5.43 Error codes in parameter RET_VAL of SFC51 *RDSYSST*

Error Code (W#16#...)	Explanation
0000	No error
0081	Length of the result field too short. (As many data records as possible are supplied anyway. Their number is indicated in the SZL header.)
7000	First call with REQ="0": No data transmission active. BUSY shows value "0".
7001	First call with REQ "1": Data transmission triggered. BUSY shows value "1".
7002	Intermediate call (REQ irrelevant): Data transmission already active. BUSY shows value "1"
8081	Length of the result field too short. (Not enough space for one data record.)
8082	SZL_ID is wrong or unknown on the CPU or in the SFC.

Continued on page 123

Table 5.43 Continued

Error Code (W#16#...)	Explanation
8083	INDEX wrong or not permitted
8085	The information is not available from the system at the moment (e.g., due to insufficient resources).
8086	Data record cannot be read due to a system error (bus, module, operating system).
8087	Data record cannot be read since the module does not exist or does not acknowledge the read job.
8088	Data record cannot be read since the actual type identifier differs from the set type identifier.
8089	Data record cannot be read since the module does not have diagnostic capability.
808A	Data type is not permitted for parameter DR (data types BOOL, BYTE, CHAR, WORD, DWORD, INT, and DINT are permitted), or bit address is not "0".
80A2	DP protocol error (Layer 2 error) (temporary error)
80A3	DP protocol error for user interface/user (temporary error)
80A4	Communication on communication bus interrupted. (Error occurs between CPU and external DP interface) (temporary error).
80C5	Distributed I/O not available (temporary error)
80C6	Data record transmission was terminated due to a higher-priority program processing level (priority class) called by the operating system.

SZL_HEADER r

The SZL_HEADER parameter has the following structure.

```
SZL_HEADER:STRUCT

        LENGTHDR:WORD

        N_DR:WORD

    END_STRUCT
```

After the read job, the LENGTHDR element contains the length in bytes of the data records read, and the N_DR element contains the number of data records in the field of the data records read.

5.4.6 Available System Status Sublists

Table 5.44 shows you the available system status sublists from which you can choose. However, depending on the type of module, you will always find a specific subset of the available sublists on a module.

Table 5.44 Available system status sublists

Sublist	SZL-ID
List of all SZL IDs of a module	W#16#xy00
Module identification	W#16#xy11
CPU characteristics	W#16#xy12
User memory areas	W#16#xy13
System areas	W#16#xy14
Block types	W#16#xy15
Priority classes	W#16#xy16
List of the permissible SDBs with no. < 1000	W#16#xy17
Maximum I/O configuration for S7-300	W#16#xy18
Status of the module LEDs	W#16#xy19
Interrupt/error assignment	W#16#xy21
Interrupt status	W#16#xy22
Priority classes	W#16#xy23
Operational states	W#16#xy24
Communication: Performance parameters	W#16#xy31
Communication: Status data	W#16#xy32
Diagnostic station list	W#16#xy33
Start information list	W#16#xy81
Start event list	W#16#xy82
Module rack/station state information	W#16#xy91
Rack/station status information	W#16#xy92
Diagnostic buffer on the CPU	W#16#xyA0
Module diagnostic information (DR0)	W#16#00B1
Module diagnostic information (DR1) through a geographical address	W#16#00B2
Module diagnostic information (DR1) through a logical address	W#16#00B3
Diagnostic data of a DP slave	W#16#00B4

5.4.7 Special Features of SFC51 *RDSYSST*

System function SFC51 is usually processed asynchronously. When SFC51 is called in diagnostic interrupt block OB82, and it has the system status list identifier (SZL-ID) W#16#00B1, W#16#00B2 or W#16#00B3 and the INDEX parameter contains the module address which caused the interrupt, then SFC51 is processed immediately. This means, in this situation it is processed synchronously.

On the CPU, each asynchronous execution of an SFC51 (Jobs with SZL_ID W#16#00B4 and W#16#4C91 and W#16#4092 and W#16#4292 and W#16#4692 and possibly W#16#00B1 and W#16#00B3) occupies resources (memory area) for job processing. If several jobs are active at the same time, it is ensured that all jobs are performed and do not affect each other.

However, only a certain number of SFC51 jobs can be active simultaneously. For the maximum number of possible SFC calls, see the performance data of the CPUs you are using. If the maximum amount of available resources is reached, an error code is transmitted by the RET_VAL parameter. If this happens, the SFC must be initiated again. SFC51 *RDSYSST* can only read one sublist at a time.

5.5 Reading and Writing Data Records/Parameters

5.5.1 Writing Dynamic Parameters with SFC55 *WR_PARM*

System function SFC55 *WR_PARM* (*WR*ite *PARa*Meter) transmits the with parameter RECORD specified data record to the addressed S7 module. The parameters which SFC55 transfers to the module do not overwrite the parameters of this module which are stored in the related SDB (*System Data Block*) on the S7 CPU.

Table 5.45 Parameters for SFC55 *WR_PARM*

Parameter	Declaration	Data Type	Memory Area	Description
REQ	INPUT	BOOL	I, Q, M, D, L, constant	REQ = "1": Request to write
IOID	INPUT	BYTE	I, Q, M, D, L, constant	Identifier of the address area: B#16#54 = I/O input B#16#55 = I/O output
LADDR	INPUT	WORD	I, Q, M, D, L, constant	The logical address set for this module in *HW Config* (here in hexadecimal format)
RECNUM	INPUT	BYTE	I, Q, M, D, L, constant	Data record number
RECORD	INPUT	ANY	I, Q, M, D, L	Data record
RET_VAL	OUTPUT	INT	I, Q, M, D, L	Return value of the SFC
BUSY	OUTPUT	BOOL	I, Q, M, D, L	BUSY = "1": The write procedure has not yet been concluded.

Make sure that the data record to be transferred is not a static data record (e.g., Data record 0). Also, if the data record to be written is kept in SDB100 to SDB129, make sure that the "static bit" is not set. The call parameters of SFC55 *WR_PARM* are listed in table 5.45.

Parameter description

IOID

Parameter *IOID* specifies the identifier of the address areas of the module which is addressed with LADDR. If the module addressed is a composite module, that is to say, if it is a module or submodule with both inputs and outputs, then you must state the area identifier of the lowest I/O address in *IOID*. If the addresses for the inputs and outputs are identical, specify B#16#54 as the identifier for the input.

LADDR

If the module you wish to address is a composite module, specify here the lower of the two addresses.

RECORD

The RECORD parameter specifies – on the CPU – the dynamic record of data type ANY-Pointer to be transferred. The data record is read with the first call of the system function. If the transmission of the data record takes longer than one CPU cycle and therefore there are repeated system function calls for the same job, then the information stored in the RECORD parameter no longer applies to these subsequent calls.

RET_VAL

The RET_VAL output parameter indicates successful or unsuccessful processing of SFC55. The error codes apply to both SFC56 and SFC57. Job processing errors (error code W#16#8xyz) that are not caused by incorrect definition of an input or output parameter of the SFC can be divided into two groups:

- Temporary errors (error codes W#16#80A2 to 80A4 and 80Cx)

  You can correct errors of this nature by calling the SFC again. The error message "W#16#80C3" is an example of a temporary error. It indicates that required resources (memory space) were being used by other functions at the time the call occurred.

- Permanent errors (error codes W#809x, 80A1, 80Bx, and 80Dx)

  Permanent errors must be corrected. You should not call the SFC again until you have seen to it that the reported error is corrected. An example of a permanent error is an incorrect length in the RECORD parameter (W#16#80B1).

Table 5.46 shows the specific error codes for SFC55, SFC56, and SFC57.

Tabelle 5.46   Error codes that apply to SFC55, SFC56 and SFC57

Error Code W#16#...	Explanation	Restrictions
7000	First call with REQ = "0": No data transmission active. BUSY has the value "0".	–
7001	First call with REQ = "1": Data transmission triggered. BUSY has the value "1".	Distributed I/O

Continued on page 127

Table 5.46 Continued

Error Code W#16#...	Explanation	Restrictions
7002	Intermediate call (REQ irrelevant): Data transmission already active. BUSY has the value "1."	Distributed I/O
8090	Specified logical base address invalid. No assignment in SDB1/SDB2x exists or no base address is specified.	–
8092	A type other than BYTE is specified in the parameter of data type ANY-Pointer.	Only for S7-400 for SFC55 *WR_PARM*
8093	This SFC is not permitted for the module selected by means of LADDR and IOID. (Permitted are S7-300 modules for S7-300, S7-400 modules for S7-400, and S7 DP modules for S7-300 and S7-400.)	–
80A1	Negative acknowledgment while the data record was being sent to the module. (Module defective or removed during data transmission).	–
80A2	DP protocol error in Layer 2. Possible hardware defect.	Distributed I/O
80A3	DP protocol error for Direct Data Link Mapper or in User Interface. Possibly a hardware defect.	Distributed I/O
80A4	Communication bus (K bus) faulty	Error occurs between CPU and external DP interface.
80B0	SFC not possible for module type since module doesn't recognize the data record	–
80B1	Incorrect length of the transferred data record.	–
80B2	The configured slot is not used.	–
80B3	Actual module type does not match the set module type in SDB1	–
80C1	The data of the preceding write job on the module for the same data record has not yet been processed by the module.	–
80C2	At the moment, the module is processing the maximum number of jobs for one CPU.	–
80C3	Required resources (memory, and so on) are busy at the moment.	–
80C4	Communication error: – Parity error – SW-Ready not set – Error in block length – Checksum error on CPU side – Checksum error on module side	–
80C5	Distributed I/O not available	Distributed I/O

Continued on page 128

127

Table 5.46 Continued

Error Code W#16#...	Explanation	Restrictions
80C6	Data record transmission was terminated due to a higher-priority program processing level (priority class) called by the operating system.	Distributed I/O
80D0	The related SDB has no entry for the module.	–
80D1	The data record number is not configured in the related SDB for the module. (Data record 241 is rejected by STEP 7.)	–
80D2	According to type identifier, the module cannot be loaded with a parameter set.	–
80D3	The SDB cannot be accessed since it does not exist.	–
80D4	Internal SDB structure error: Internal SDB structure management pointer is pointing to an area outside the SDB.	Only for S7-300
80D5	Data record is static.	Only for SFC55 WR_PARM

Notes on calling SFC51 in S7-400 systems

- It could happen that the general error code W#16#8544 occurs. This only means that access to at least one byte of the I/O memory area contained in the data record was blocked, but data transmission was continued anyway.

- It could happen that system functions SFC55 to SFC59 return the error code W#16#80Fx. This means that an error has occurred which cannot be precisely localized.

5.5.2 Writing a Predefined Data Record/Parameter from the SDB with SFC56 *WR_DPARM*

System function SFC56 *WR_DPARM* (*WR*rite *D*efault *PARaM*eter) transfers the static or dynamic data record with the number RECNUM from the SDB of the S7 CPU to the module addressed by LADDR and IOID. In S7-300 systems, the system data block can be in the range from SDB100 to SDB103, whereas S7-400 systems use the range SDB100 to SDB129 for this purpose. Table 5.47 shows the input and output parameters of SFC56 *WR_DPARM*.

Parameter description

IOID

The parameter specifies the identifier of the address area of the module which is addressed with LADDR. If the module addressed is a composite module, that is to say, if it is a module or submodule with both inputs and outputs, then you must state the area identifier of the lowest I/O address in *IOID*. If the addresses for the inputs and outputs are identical, specify B#16#54 as the identifier for the input.

Table 5.47 Parameters of SFC56 *WR_DPARM*

Parameter	Declaration	Data Type	Memory Area	Description
REQ	INPUT	BOOL	I, Q, M, D, L, constant	REQ = "1": Request to write
IOID	INPUT	BYTE	I, Q, M, D, L, constant	Identifier of the address area: B#16#54 = I/O input B#16#55 = I/O output
LADDR	INPUT	WORD	I, Q, M, D, L, constant	Logical address set for this module in *HW Config* (here in hexadecimal format)
RECNUM	INPUT	BYTE	I, Q, M, D, L, constant	Data record number
RET_VAL	OUTPUT	INT	I, Q, M, D, L	Return value of the SFC
BUSY	OUTPUT	BOOL	I, Q, M, D, L	BUSY = "1": The write procedure is not yet concluded.

LADDR

If the module you wish to address is a composite module, specify here the lower of the two addresses.

RET_VAL

The error codes for SFC56 are the same as those of the RET_VAL values of SFC55 shown in table 5.46.

5.5.3 Writing All Predefined Data Records/Parameters from the SDB with SFC57 *PARM_MOD*

SFC57 *PARM_MOD* (*PAR*a*M*eterize *MOD*ule) transfers all static or dynamic data records of a module from the associated system data block (SDB) to the addressed module. You will have defined this SDB in the *HW Config* program. In S7-300 systems, the SDB lies in the range from SDB100 to SDB103, whereas S7-400 systems use the range SDB100 to SDB129 for this purpose. Table 5.48 shows the input and output parameters of SFC57 *PARM_MOD*.

Parameter description

IOID

This parameter specifies the identifier of the address area of the module which is addressed with LADDR. If the module addressed is a composite module, that is to say, if it is a module or submodule with both inputs and outputs, then you must state the area identifier of the lowest I/O address in *IOID*. If the addresses for the inputs and outputs are identical, specify B#16#54 as the identifier for the input.

LADDR

If the module you wish to address is a composite module, specify here the lower of the two addresses.

Table 5.48   Parameters of SFC57 *PARM_MOD*

Parameter	Declaration	Data Type	Memory Area	Description
REQ	INPUT	BOOL	I, Q, M, D, L, constant	REQ = "1": Request to write
IOID	INPUT	BYTE	I, Q, M, D, L, constant	Identifier of the address area: B#16#54 = I/O input (PE) B#16#55 = I/O output (PA)
LADDR	INPUT	WORD	I, Q, M, D, L, constant	Logical address set for this module in *HW Config* (here in hexadecimal format)
RET_VAL	OUTPUT	INT	I, Q, M, D, L	Return value of the SFC
BUSY	OUTPUT	BOOL	I, Q, M, D, L	BUSY = "1": The write procedure is not yet concluded.

RET_VAL parameter

The error codes for SFC57 are the same as the RET_VAL values for SFC55 shown in table 5.46.

5.5.4   Writing a Data Record/Parameter with SFC58 *WR_REC*

SFC58 *WR_REC* (*WR*ite *REC*ord) transfers the data record specified in the RECORD parameter to the module addressed with LADDR and IOID. In contrast to SFC55, SFC58 can only be used to transfer data records with the numbers 2 to 240. Table 5.49 lists the input and output parameters of SFC58 *WR_REC*.

Table 5.49   Parameters of SFC58 *WR_REC*

Parameter	Declaration	Data Type	Memory Area	Description
REQ	INPUT	BOOL	I, Q, M, D, L, constant	REQ = "1": Request to write
IOID	INPUT	BYTE	I, Q, M, D, L, constant	Identifier of the address area: B#16#54 = I/O input B#16#55 = I/O output
LADDR	INPUT	WORD	I, Q, M, D, L, constant	Logical address set for this module in *HW Config* (here in hexadecimal format)
RECNUM	INPUT	BYTE	I, Q, M, D, L, constant	Data record number (permissible values: 2 to 240)
RECORD	INPUT	ANY	I, Q, M, D, L	Data record – only data type BYTE is permitted.
RET_VAL	OUTPUT	INT	I, Q, M, D, L	Return value of the SFC
BUSY	OUTPUT	BOOL	I, Q, M, D, L	BUSY = "1": The write procedure is not yet concluded.

Parameter description

IOID

This parameter specifies the identifier of the address area of the module which is addressed with LADDR. If the module addressed is a composite module, that is to say, if it is a module or submodule with both inputs and outputs, then you must state the area identifier of the lowest I/O address in *IOID*. If the addresses for the inputs and outputs are identical, specify B#16#54 as the identifier for the input.

LADDR

If the module you wish to address is a composite module, specify here the lower of the two addresses.

RET_VAL

The error codes transmitted by the RET_VAL parameter are listed in table 5.50.

Table 5.50   Error codes that apply to SFC58 *WR_REC*

Error Code W#16#...	Explanation	Restrictions
7000	First call with REQ = "0": No data transmission active. BUSY has the value "0".	–
7001	First call with REQ = "1": Data transmission triggered. BUSY has the value "1".	Distributed I/O
7002	Intermediate call (REQ irrelevant): Data transmission already active. BUSY has the value "1".	Distributed I/O
8090	Specified logical base address invalid: No assignment in SDB1/SDB2x exists for the specified logical base address, or no base address was specified when the function was called.	–
8092	A type other than BYTE is specified in the parameter of data type ANY-Pointer.	Only for S7-400
8093	This SFC is not permitted for the module selected by means of LADDR and IOID. (Permissible are S7-300 modules for S7-300, S7-400 modules for S7-400, and S7 DP modules for S7-300 and S7-400.)	–
80A0	Negative acknowledgment while reading from a module. (Module defective or removed during the read-access.)	Only for SFC59 *RD_REC*
80A1	Negative acknowledgment while writing to a module. (Module defective or removed during the write-access.)	Only for SFC58 *WR_REC*
80A2	DP protocol error in Layer 2. Possible hardware defect	Distributed I/O
80A3	DP protocol error during Dirket Data Link Mapper or in User Interface. Possibly a hardware defect.	Distributed I/O

Continued on page 132

Table 5.50 Continued

Error Code W#16#...	Explanation	Restriction
80A4	Communication on K bus faulty	Error occurs between CPU and external DP interface.
80B0	Possible cause: – SFC call for this module type not possible – Data record unknown to module – Data record numbers greater than 240 are not permitted. – With SFC58 *WR_REC*, data records 0 and 1 are not permitted.	–
80B1	Length specification in RECORD parameter is wrong: – For SFC58 *WR_REC*: Data record length is wrong. – For SFC59 *RD_REC* (only possible when older S7-300 FMs and S7-300 CPs are used): specification > data record length – With SFC13 *DPNRM_DG*: Specification < data record length	–
80B2	The configured slot is not occupied.	–
80B3	Actual module type does not match set module type in SDB1	–
80C0	With – SFC59 *RD_REC*: Although the module maintains the data record, there is no data yet to be read. – SFC13 *DPNRM_DG*: No diagnostic data available.	Only for SFC59 *RD_REC* or for SFC13 *DPNRM_DG*
80C1	The data of the preceding write job on the module for the same data record have not yet been processed by the module.	–
80C2	At the moment, the module is processing the maximum number of jobs for one CPU.	–
80C3	Required resources (e.g., memory and so on) are busy at the moment.	–
80C4	Internal communication error: – Parity error – SW-Ready not set – Error in block length – Checksum error on CPU side – Checksum error on module side	
80C5	Distributed I/O is not available.	Distributed I/O
80C6	The data record transmission was terminated due to a higher-priority program processing level (priority class) called by the operating system.	Distributed I/O

In S7-400 systems it can happen that SFC58 returns error code W#16#80Fx. This means that an error has occurred which cannot be precisely localized.

5.5.5 *Reading Data Record with SFC59 RD_REC*

SFC59 *RD_REC* (*ReaD RECord*) reads data record RECNUM (area 0 to 240) from the addressed module and stores it in the destination area specified by the RECORD parameter. Table 5.51 lists the input and output parameters of SFC59 *RD_REC*.

Table 5.51 Parameters for SFC59 *RD_REC*

Parameter	Declaration	Data Type	Memory Area	Description
REQ	INPUT	BOOL	I, Q, M, D, L, constant	REQ = "1": Request to write
IOID	INPUT	BYTE	I, Q, M, D, L, constant	Identifier of the address area: B#16#54 = I/O input B#16#55 = I/O output
LADDR	INPUT	WORD	I, Q, M, D, L, constant	Logical address set for this module in *HW Config*, here in hexadecimal format
RECNUM	INPUT	BYTE	I, Q, M, D, L, constant	Data record number (permissible values: 0 to 240)
RET_VAL	OUTPUT	INT	I, Q, M, D, L	Error code
BUSY	OUTPUT	BOOL	I, Q, M, D, L	BUSY = "1": The read procedure is not yet concluded.
RECORD	OUTPUT	ANY	I, Q, M, D, L	Destination area for the data record read

Parameter description

IOID

This parameter specifies the identifier of the address area of the module which is addressed with LADDR. If the module addressed is a composite module, that is to say, if it is a module or submodule with both inputs and outputs, then you must state the area identifier of the lowest I/O address in *IOID*. If the addresses for the inputs and outputs are identical, specify B#16#54 as the identifier for the input.

LADDR

If the module you wish to address is a composite module, specify here the lower of the two addresses.

RET_VAL

If an error occurs while the function is being executed, the RET_VAL parameter transmits an error code. The error codes are the same as those of SFC58 which are listed in table 5.50. In S7-400 systems, SFC59 can also return error code W#16#80Fx. This means that an error has occurred which cannot be precisely localized.

RECORD

The RECORD output parameter specifies the length of the data record to be read from the selected data record. This means that the length you specify here must not be greater than the actual length of the data record. For this reason, make sure that the length you specify in RECORD is exactly the same as the length of the actual data record to be read.

In addition, when SFC59 is processed asynchronously, remember that the RECORD parameter contains the same length information for all subsequent calls. Remember to only use data type BYTE.

5.5.6 Read Data Set with SFB52 *RDREC*

With the SFB52 "RDREC" (ReaD RECord) you read a data record with the number INDEX from a DP slave component (component or module) that has been adressed via ID. The data read are stored in the area defined by RECORD.

The parameter MLEN defines the maximum number of bytes you want to read. A minimum length of MLEN bytes must therefore be selected for the destination area RECORD.

If an error occurred during transfer of the data set, this is indicated via the ERROR output parameter. The STATUS output parameter contains the error information in this case.

The interface of SFB52 "RDREC" is identical with that of the FB "RDREC" defined in the PNO AK 1131 standard. The input and output parameters of SFB52 *RDREC* are shown in Table 5.52.

Table 5.52   Parameter for SFB52 *RDREC*

Parameter	Declaration	Data type	Memory area	Description
REQ	INPUT	BOOL	I, Q, M, D, L, constant	REQ = "1": write request
ID	INPUT	DWORD	I, Q, M, D, L, constant	Logical address of the DP slave component (module)
INDEX	INPUT	INT	I, Q, M, D, L, constant	Data set number
MLEN	INPUT	INT	I, Q, M, D, L, constant	Maximum length in bytes of the data set information to be read
VALID	OUTPUT	BOOL	I, Q, M, D, L	New data record has been received and is valid
BUSY	OUTPUT	BOOL	I, Q, M, D, L	BUSY = 1: write operation not yet completed
ERROR	OUTPUT	BOOL	I, Q, M, D, L	ERROR = 1: error occurred during write operation

Continued on page 135

Table 5.52 Continued

Parameter	Declaration	Data type	Memory area	Description
STATUS	OUTPUT	DWORD	I, Q, M, D, L	Call ID in byte 2 and 3 (W#16#7001 or W#16#7002) or the error code
LEN	OUTPUT	INT	I, Q, M, D, L	Length of the read data set information
RECORD	IN_OUT	ANY	I, Q, M, D, L	Data set

Description of parameters

VALID

The value TRUE of the VALID output parameter indicates that the data set has been successfully transferred to the destination area RECORD. In this case, the LEN output parameter contains the length of the read data in bytes.

RECORD

Due to the asynchronous processing of SFB52, care must be taken that the RECORD actual parameter has the same value in all calls belonging to one and the same job.

STATUS

The STATUS output parameter contains error information. The parameter is explained in detail at the end of Section 5.5.7.

5.5.7 Write Data Set with SFB53 *WDREC*

Data set RECORD with the number INDEX (range 0 to 255) is transferred to the component (module) of a DP slave addressed by ID using SFB53 *WRREC* (*WR*ite *REC*ord).

The length in bytes of the data set to be transferred is defined with the LEN parameter. For this reason, a minimum of LEN bytes must be selected for the RECORD source area.

If an error occurred during transfer of the data set, this is indicated via the ERROR output parameter. The STATUS output parameter contains the error information in this case.

The interface of SFB53 *WRREC* is identical with that of the FB "WRREC" defined in the PNO AK 1131 standard. The input and output parameters of SFB53 *WRREC* are shown in Table 5.53.

Table 5.53 Parameters for SFB53 *WRREC*

Parameter	Declaration	Data type	Memory area	Description
REQ	INPUT	BOOL	I, Q, M, D, L, constant	REQ = "1": write request
ID	INPUT	DWORD	I, Q, M, D, L, constant	Logical address of the DP slave component (module)
INDEX	INPUT	INT	I, Q, M, D, L, constant	Data set number
LEN	INPUT	INT	I, Q, M, D, L, constant	Maximum length in bytes of the data set to be transferred
DONE	OUTPUT	BOOL	I, Q, M, D, L	Data set has been transferred
BUSY	OUTPUT	BOOL	I, Q, M, D, L	BUSY = 1: write operation not yet completed
ERROR	OUTPUT	BOOL	I, Q, M, D, L	ERROR = 1: error occurred during write operation
STATUS	OUTPUT	DWORD	I, Q, M, D, L	Call ID in byte 2 and 3 (W#16#7001 or W#16#7002) or the error code
RECORD	IN_OUT	ANY	I, Q, M, D, L	Data set

Description of parameters

DONE

The value TRUE of the DONE output parameter indicates that the data set has been successfully transferred to the DP slave. Transfer of the data set has been completed when the BUSY output parameter has assumed the value FALSE.

RECORD

Due to the asynchronous processing of SFB53, care must be taken that the RECORD actual parameter has the same value in all calls belonging to one and the same job. The same applies to the LEN actual parameter.

STATUS

The STATUS output parameter contains error information. If it is interpreted as ARRAY[1...4] OF BYTE, the error information has the structure shown in Table 5.54.

The status of the job is indicated in conjunction with the BUSY output parameter and bytes 2 and 3 of the STATUS output parameter. Bytes 2 and 3 of STATUS correspond here to the RET_VAL output parameter of the asynchronously operating SFCs (also see Table 5.50).

Table 5.54 Representation of the STATUS output parameter

Field element	Name	Meaning
STATUS[1]	Function_Num	B#16#00, if no error Function ID from DPV1-PDU: In the event of a fault, B#16#80 is ORed up. If no DPV1 protocol element is used: B#16#C0.
STATUS[2]	Error_Decode	Location of fault ID
STATUS[3]	Error_Code_1	Error code
STATUS[4]	Error_Code_2	Vendor-specific expansion of the error code

The location of the error code is entered in STATUS[2] and shown in Table 5.55. The error code itself is in STATUS[3] and is represented in Table 5.56.

The error code in STATUS[4] is passed from the DP master to the CPU in the case of DPV1 errors. If there is no DPV1 error, the value is set to "0", with the following exceptions in the case of SFB52:

- STATUS[4] contains the length of the destination area from RECORD, if MLEN is > the length of the destination area from RECORD

- STATUS[4]=MLEN, if the actual data set length is < MLEN < the length of the destination area from RECORD

Table 5.55 Error codes in STATUS[2]

Error_Decode (B#16#...)	Source	Meaning
00 to 7F	CPU	No error, or no warning
80	DPV1	Error in accordance with IEC 61158-6
81 to 8F	CPU	B#16#8x shows an error in the nth call parameter of the SFB
FE, FF	DP Profile	Profile-specific error

Table 5. 56 Error codes in STATUS[3]

STATUS[2] Error_Decode (B#16#...)	STATUS[3] Error_Code_1 (B#16#...)	Explanation according to DVP1	Meaning
00	00		No error, no warning
70	00	Reserved, reject	Initial call; no data set transfer active
	01	Reserved, reject	Initial call; data set transfer initiated
	02	Reserved, reject	Intermediate call; data set transfer already active
80	90	Reserved, pass	Logical starting address invalid

Continued on page 138

137

Table 5. 56 Continued

STATUS[2] Error_Decode (B#16#...)	STATUS[3] Error_Code_1 (B#16#...)	Explanation according to DVP1	Meaning
	92	Reserved, pass	Impermissible type at ANY pointer
	93	Reserved, pass	The DP component addressed using ID or F_ID is not configured
	A0	Read error	Negative acknowledgement when reading from the module
	A1	Write error	Negative acknowledgement when writing to module
	A2	Module failure	DP protocol error on Layer 2, possible hardware defect
	A3	Reserved, pass	DP protocol error on direct data link mapper or user-interface/user, possible hardware defect
	A4	Reserved, pass	Communication fault on K bus
	A5	Reserved, pass	–
	A7	Reserved, pass	DP resource in use
	A8	Version conflict	Version conflict
	A9	Feature not supported	Feature not supported
	AA to AF	User specific	Specific to DP master
	B0	Invalid index	Module does not recognize data set: data set number >= 256 not permissible
	B1	Write length error	Length error in AINFO
	B2	Invalid slot	The configured slot is not occupied
	B3	Type conflict	Actual module type not identical to desired module type
	B4	Invalid area	Invalid range
	B5	State conflict	State conflict
	B6	Access denied	Access denied
	B7	Invalid range	Invalid range
	B8	Invalid parameter	Invalid parameter
	B9	Invalid type	Invalid type
	BA to BF	User specific	Specific to DP master
	C0	Read constrain conflict	The module has the data set but there are no read data yet

Continued on page 139

Table 5. 56 Continued

STATUS[2] Error_Decode (B#16#...)	STATUS[3] Error_Code_1 (B#16#...)	Explanation according to DVP1	Meaning
	C1	Write constrain conflict	The data of the previous write job on the module for the same data set have not yet been processed by the module
	C2	Resource busy	The module is currently processing the possible maximum of jobs for a CPU
	C3	Resource unavailable	The required resource is currently in use
	Dx	User specific	Specific to DP slave. See description of DP slave
81	00 to FF		Error in first call parameter (in case of SFB54: MODE)
	00		Impermissible mode
82	00 to FF		Error in second call parameter
↓	↓		↓
88	00 to FF		Error in eighth call parameter (in case of SFB54: TINFO)
	01		Wrong syntax ID
	23		Quantity structure exceeded or destination area too small
	24		Wrong area ID
	32		DB/DI no. outside user range
	3A		DB/DI no. is ZERO in the case of area code DB/DI, or specified DB/DI is not available
89	00 to FF		Error in ninth call parameter (in case of SFB54: AINFO)
	01		Wrong syntax ID
	23		Quantity structure exceeded or destination area too small
	24		Wrong area ID
	32		DB/DI no. outside user range
	3A		DB/DI no. is ZERO in the case of area code DB/DI, or specified DB/DI is not available
8A	00 to FF		Error in 10th call parameter

Continued on page 140

Table 5. 56 Continued

STATUS[2] Error_Decode (B#16#...)	STATUS[3] Error_Code_1 (B#16#...)	Explanation according to DVP1	Meaning
↓	↓		↓
8F	00 to FF		Error in 15th call parameter
FE, FF	00 to FF		Profile-specific error

6 Example Projects for Data Communication with PROFIBUS DP

Introduction

SIMATIC S7 systems handle the distributed input/output peripherals connected by a PROFIBUS DP network in the same way as the I/O modules connected locally in the central rack or expansion rack. Depending on the addresses you allocated during hardware configuration in the *HW Config* program, the input/output data is exchanged either directly through the process image, or by means of I/O access commands.

- SIMATIC S7 systems provide system functions SFC14 *DPRD_DAT* and SFC15 *DPWR_DAT* for data communication with complex DP slaves that have a consistent input/output data area.

- An S7-300 programmable controller that acts as an I-slave and is equipped with a CPU315-2DP can trigger a process interrupt on the DP master by means of system function SFC7 *DP_PRAL*.

- Module parameter data of S7 DP slaves can be read or written from within the user program. System functions are provided for this purpose.

- Activation of output signals and acquisition of input signals on DP slaves can be synchronized using system function SFC11 *DPSYC_FR*.

This chapter gives you some practical examples of data communication with DP slaves in SIMATIC S7 systems. The hardware of the example projects is the same as in the example configured earlier in chapter 4. The example programs are shown as alphanumeric program lists in STL presentation. You should therefore have a basic knowledge of Statement List programming (STL).

6.1 Data Communication with I/O Access Commands

The CPUs of SIMATIC S7 systems address I/O data of distributed peripheral modules by means of special I/O access commands programmed in the STEP 7 program. These commands invoke direct I/O access or I/O access through the process image. The data format for reading and writing distributed I/O information can be bytes, words or double words. Figure 6.1 illustrates I/O communication with DP slaves in different data formats.

However, some DP slave modules have more complex data structures. Their input and output data areas have a length of 3 bytes or more than 4 bytes. These are so-called consistent I/O data areas. In the parameter set for DP slaves using consistent data areas, the parameter "consistency" must be set to "Total length" (see also section 2.2.2 on configu-

CPU	I/O area	DP master	DP slave
Byte access	Byte n	Byte n	Byte n
Word access	Byte n / Byte n+1	Byte n / Byte n+1	Byte n / Byte n+1
Double word access	Byte n / Byte n+1 / Byte n+2 / Byte n+3	Byte n / Byte n+1 / Byte n+2 / Byte n+3	Byte n / Byte n+1 / Byte n+2 / Byte n+3

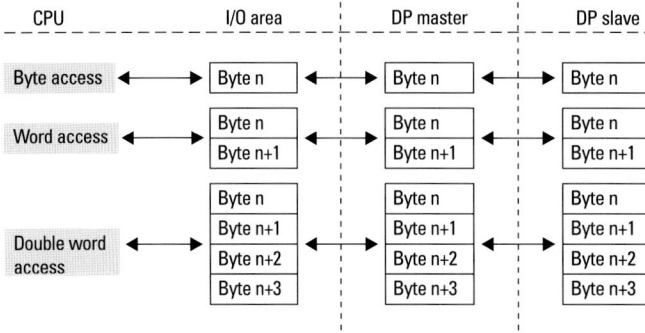

Figure 6.1
Input/output data communication with DP slaves using STEP 7 and
I/O access commands

ration data). With consistent data, the input and output data can no longer pass through the process image, nor can the data exchange be invoked by normal I/O access commands. The reason for this lies in the CPU update cycle for the input/output data on the DP master. The update of DP input/output data is determined exclusively by the cyclic data exchange (bus cycle) of the DP master with the DP slaves (see figure 6.2). Therefore, the data to and from the DP master may have already changed between one STEP 7 access instruction addressing DP slave I/O data and the next I/O access instruction. Due to this, data consistency can only be ensured for those I/O structures and areas that the user program can address without any interruption using byte, word or double-word commands.

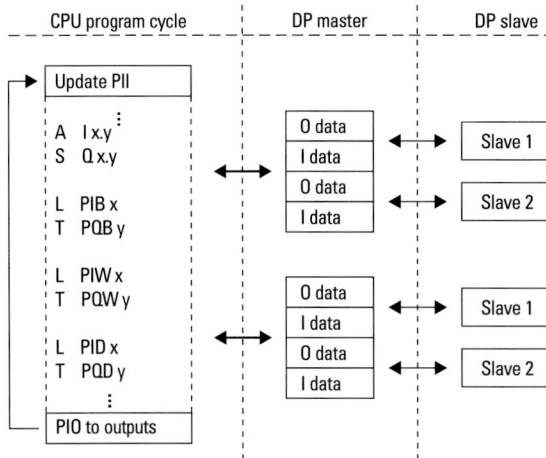

CPU program cycle | DP master | DP slave

Update PII

A I x.y
S Q x.y

O data / I data — Slave 1
O data / I data — Slave 2

L PIB x
T PQB y

L PIW x
T PQW y

O data / I data — Slave 1
O data / I data — Slave 2

L PID x
T PQD y

PIO to outputs

Update PII (process image input table). Transfer PIO (process image output table) to outputs.

Figure 6.2   Input/output data of the DP slaves. Update and access.

6.2 Exchanging Consistent Data with SFC14 *DPRD_DAT* and SFC15 *DPWR_DAT*

DP slaves that have to control complex functions, such as closed-loop controllers or electrical drives, can generally not use simple data structures for these tasks. These DP slaves require larger input and output areas. The information in these I/O areas is usually coherent and must not be separated. It can therefore not be stored in a byte, word or double-word structure. Such a data area is said to be "consistent" (see also section 2.2.2 on configuration data). Within one input/output module, data areas with a length of up to 64 bytes or words (128 bytes) can be specified using the configuration telegram. To read or write information to and from these module-related, consistent input/output data areas of the DP slaves, special functions must be used. In SIMATIC S7, system functions SFC14 *DPRD_DAT* and SFC15 *DPWR_DAT* are reserved for this purpose.

Figure 6.3 shows the principle of operation of system functions SFC14 and SFC15. The LADDR parameter of the SFC is a pointer to the input data area to be read from, or to the output data area to be written to. In the SCF parameter you must specify the same start address of the DP slave input or output module which you have already configured earlier in the *HW Config* program; but this time specify the address in hexadecimal format. The RECORD parameter of the SFC defines the respective source or destination area for the data on the CPU. For a description of the input and output parameters and the values (error codes) returned by the RET_VAL parameter, see section 5.3.

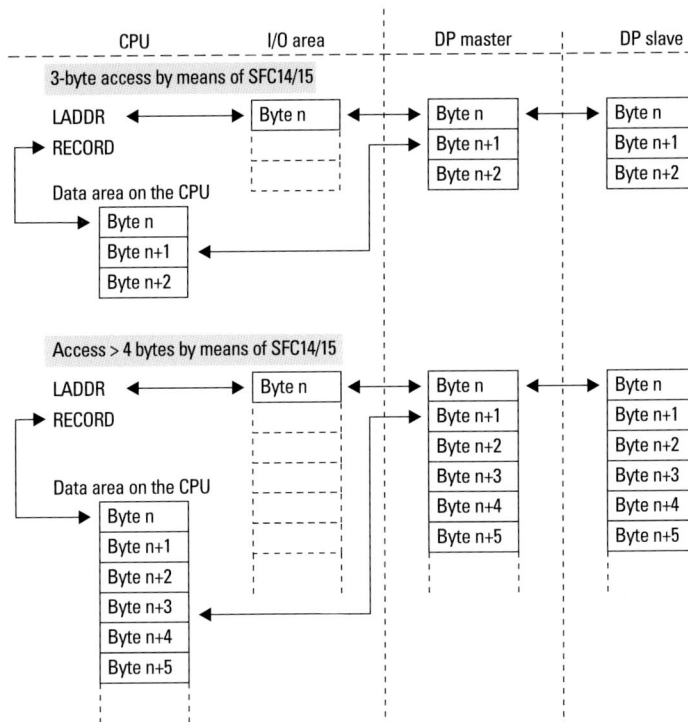

Figure 6.3 Input/output data of the DP slaves using SFC14 and SFC15

The following example project illustrates the use of SFC14 and SFC15. We will use the same hardware as described in section 4.2.5 (under the heading <u>S7-300/CPU315-2DP as I slave</u>). However, we will restrict our project to one S7 DP master station (S7-400) and one I slave (S7-300). Therefore, you will have to delete the ET200B and ET200M nodes which you configured earlier, in the example given in chapter 4. Connect the DP interfaces of the S7-300 and S7-400 controllers with each other using an appropriate PROFIBUS cable, and switch on the power supply of the devices. Our project is based on the assumption that both programmable controllers have been reset. That is to say, their work memory, load memory and system memory are completely cleared. Both PLCs are in the RUN mode (switch position RUN-P).

The two consistent input/output data areas for the I slave have a length of 10 bytes each and their parameter "Consistency" is set to "Total length" (see also section 4.2.5, figure 4.18). This means that you will have to use the system functions SFC14 and SFC15 for input/output data communication on the I slave and on the DP master.

6.2.1 User Program for I-Slave (S7-300 with CPU315-2DP)

The I slave of our example project has consistent input/output areas of more than 4 bytes. Therefore, just as with the S7 DP master, you must use the system functions SFC14 and SFC15 to transmit the I/O data. Remember that the output data sent by means of SFC15 on the DP master is read by SFC14 on the I-slave and treated as input data. The reverse applies for the input data of the DP master coming from the I-slave. This is illustrated in figure 6.4.

The CPUs of the SIMATIC S7-300 controller do not recognize addressing errors. Therefore, on the CPU315-2DP, you may place the I/O data to be transmitted by the SFCs in a process image area which is not allocated otherwise. You may for instance use IB100 to IB109 and QB100 to QB 109. In the user program, you can thus address this data with simple bit, byte, word and double-word instructions.

Now, let's generate the user program required for the I slave.

- In *SIMATIC Manager,* open the project S7_PROFIBUS_DP. Double-click folder *SIMATIC 300(1),* and go through the folders *CPU315-2DP* and *S7 Program* to reach the *Blocks* folder (figure 6.5). Double-click the *Blocks* folder to open it. Organization block OB1 and the system data blocks (SDBs) generated in *HW Config* are already set up in this folder. Note that you must have saved and compiled your hardware configuration in *HW Config;* otherwise *HW Config* cannot generate the system data and you will then see no *System Data* folder at this point.

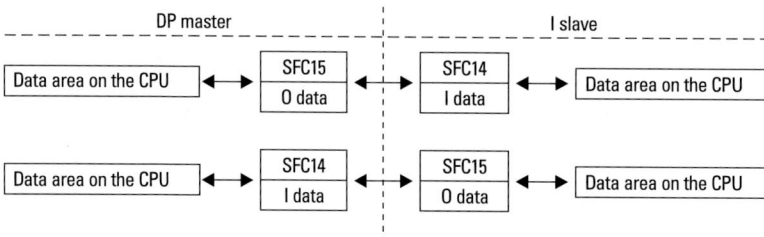

Figure 6.4 I/O data communication with I slave by means of SFC14 and SFC15 (example project)

- Double-click *OB1* to open it. This automatically opens the STEP 7 program *LAD/STL/ FBD*. Now, write the program for OB1 in the STL view.

- In the program editor, enter the command "CALL SFC14", and confirm with the RETURN key. SFC14 *DPRD_DAT* appears with its input and output parameters. Define the input and output parameters as shown in figure 6.6. Enter the load and transfer instructions. Then call SFC15, and again define the input and output parameters for this system function. When these two SFCs are called, the related block shells for these standard functions are automatically copied from the STEP 7 standard library *(...\SIEMENS\STEP7\S7libs\STDLIB30)* to the folder named *Blocks*.

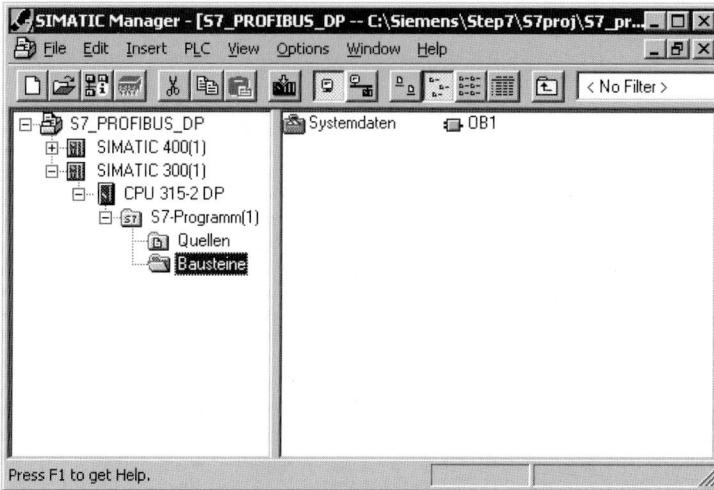

Figure 6.5   *SIMATIC Manager* with the *Blocks* folder opened

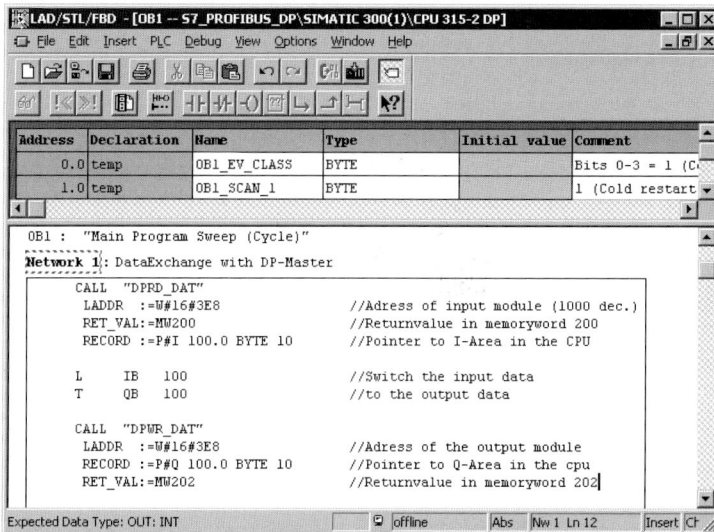

Figure 6.6   STL program editor in STEP 7 with OB1 (example program for the CPU315-2DP)

- To be able to easily monitor data communication on the DP master, route the first data byte received (IB100) to the first byte to be sent (QB100) by using appropriate load and transfer instructions (see figure 6.6). This will later copy the first data byte sent by the DP master from the input data area of the I slave immediately back to the output data area of the I slave, and thus back to the DP master.

- Use SAVE to save OB1, and close the program editor (the STL editor in our case) for OB1. In the task bar of Windows 95/NT, change to *SIMATIC Manager*. The *Blocks* folder should now contain the block objects *System data*, *OB1*, *SFC14* and *SFC15*.

When the DP master changes its operating mode or breaks down, the operating system reacts by calling up certain OBs on the I-slave. Were these OBs missing on the I-slave, the CPU would automatically switch to STOP. The next step is therefore to set up the relevant error OBs on the I slave to prevent the CPU from going into STOP in these situations.

- When the CPU of the DP master changes from RUN to STOP, organization block OB82 (diagnostic interrupt) is called on the I-slave. To prevent a CPU STOP due to a nonexistent OB82, insert OB82 in the *Blocks* folder of the *SIMATIC 300(1)* station. Proceed as follows: right-click on the *Blocks* folder to open the context menu, then select INSERT NEW OBJECT → ORGANIZATION BLOCK. In the "Properties – Organization Block" dialog box, type in OB82 in the "Name" field, and quit this dialog box with OK. Back in *SIMATIC Manager*, you can see that object *OB82* has now been inserted in the *Blocks* folder.

- When the DP master breaks down, organization block OB86 (rack failure) is called on the I slave. To prevent the CPU of the I slave from going to STOP in such a situation, you must set up OB86. Proceed as described for OB82. Use and evaluation of these error OBs is described in detail in section 7.

- Use the DOWNLOAD button from the toolbar, or select PLC → DOWNLOAD from the menu bar, to copy all blocks from the *Blocks* folder to the CPU315-2DP. For this, you must have connected your PG programming unit or PC to the CPU315-2DP using the *MPI* cable, and the power supply of the PLC must be on. During download, the operating mode switch of the CPU315-2DP must be in the RUN-P or STOP position.

- After download, switch the CPU315-2DP back to RUN. This means, if the operating mode switch was in STOP during download, change it now from STOP to RUN-P. If the switch was already in the RUN-P position during download, you will automatically be asked whether the CPU315-2DP should be started now. Confirm with OK. The LEDs of the CPU315-2DP for the DP interface have the following status: the "SF DP" LED is on, and the "BUSF" LED is flashing.

6.2.2 User Program for DP Master (S7-400 with CPU416-2DP)

To generate the DP master program for the example project, open the *Blocks* folder of the *SIMATIC 400(1)* station. Open OB1, and call SFC14 and SFC15 as shown in figure 6.7.

```
CALL SFC14
    LADDR   : =W#16#3E8              //Start address of the input module (1000dec)
    RET_VAL : =MW200                 //Return value in memory word 200
    RECORD  : =P#DB10.DBX 0.0 BYTE 10 //Pointer to data area for input data

CALL SFC15
    LADDR   : =W#16#3E8              //Start address of the output module (1000dec)
    RECORD  : =P#DB20.DBX 0.0 BYTE 10 //Pointer to data area for output data
    RET_VAL : =MW202                 //Return value in memory word 202
```

Figure 6.7 DP master program for data communication by means of SFC14 and SFC15 (example)

To prevent a CPU STOP on the DP master due to nonexistent diagnostic and error OBs, set up OB82 and OB86 on the DP master CPU. Use data blocks DB10 and DB20 as the data areas for the input/output data of the I slave. Make sure to reserve sufficient space for these DBs.

- Select the *Blocks* folder, open the shortcut menu and use INSERT NEW OBJECT → DATA BLOCK to insert a new data block. In the "Properties – Data Block" dialog box, enter DB10 in the "Name" field, and exit this dialog box with OK.

- Double-click DB10 in the *Blocks* folder. In the "New Data Block" dialog box, select "Data block" in the *Create* group. Confirm with OK. This opens the DB Editor. Enter a BYTE-ARRAY (ARRAY = summary of elements of the same data type) with a length of 10 bytes (bytes 0 to 9) (see figure 6.8). Save DB10. Set up DB20 in the same manner, but this time type in "Outputdata" in the *Name* column. Save DB20 and close the edit screen for DB10 and DB20.

- Use the taskbar to return to the *Blocks* folder in *SIMATIC Manager*. Now, select the DOWNLOAD command to copy all blocks from the *Blocks* folder to the CPU 416-2DP. Make sure that the *MPI* cable is connected between your PG programming unit or PC and the CPU416-2DP, and that the operating mode switch of the CPU is in the STOP state.

- After download, set the operating mode switch of the CPU to RUN-P. The CPU 416-2DP must now be in RUN mode. There must be no DP-related error LEDs ("SF DP" LED or "BUSF" LED) on or flashing. If these LEDs are off, DP data communication between the DP master and the I slave will be executed without any errors.

Address	Name	Type	Initial value	Comment
*0.0		STRUCT		
+0.0	Inputdata	ARRAY[1..10]		
*1.0		BYTE		
=10.0		END_STRUCT		

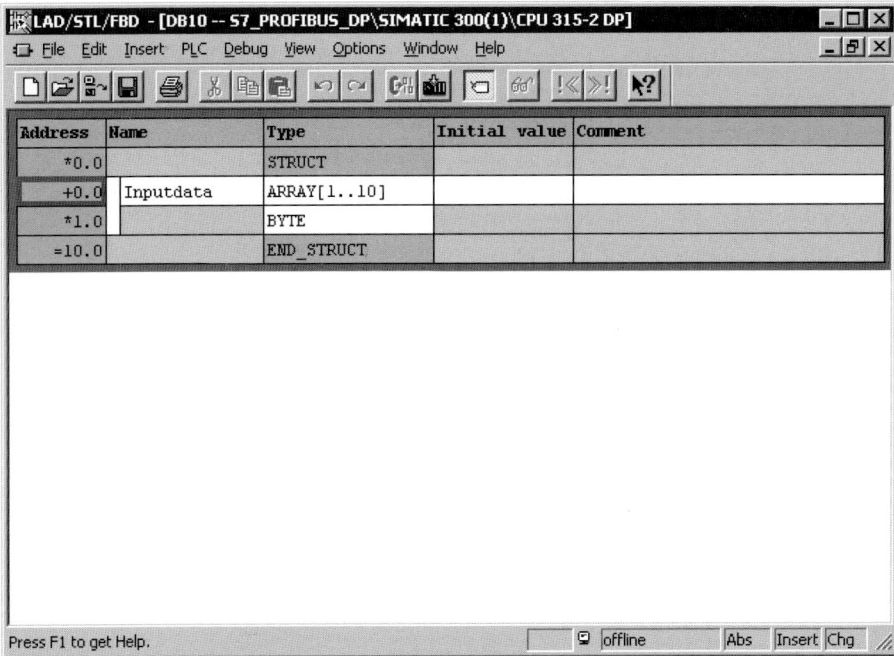

Figure 6.8   DB editor with DB10 (example program for CPU 416-2DP)

6.2.3 Testing the Data Exchange between DP Master and I Slave

To test the exchange of input/output data, select the online view for the project. In *SIMATIC Manager*, select *View* → ONLINE. Again, make sure that the MPI cable between your PG/PC programming unit and the CPU 416-2DP is properly connected.

Open the *SIMATIC 400(1)* folder and right-click *CPU 416-2DP* to open the shortcut menu. Select PLC → MONITOR/MODIFY VARIABLES. You can now change the variables and monitor the system's response.

Enter the two variables DB20.DBB0 (1st output data byte of the I slave) and DB10.DBB0 (1st input data byte of the I slave) as shown in figure 6.9. Enter a Modify Value, for instance "B#16#11", for the 1st output data byte. Now, start monitoring the variables by selecting VARIABLE → MONITOR from the menu bar, or the MONITOR (ACCORDING TO TRIGGER) button from the toolbar. The two monitor values indicate "B#16#00". Now, in the menu bar, select VARIABLE → ACTIVATE MODIFY VALUES to activate the manually entered value for the 1st output data byte of the I slave. You will see that the monitor values of both variables immediately change to the set value. The reason for this is that the data that the I slave received from the DP master is immediately returned to the DP master by the user program.

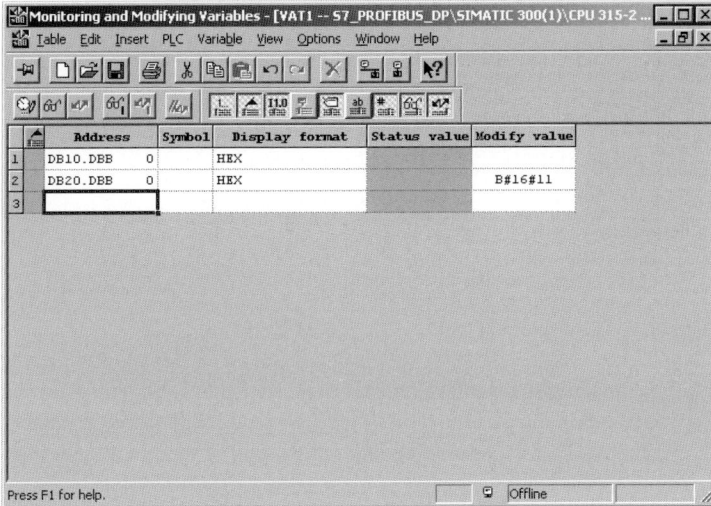

Figure 6.9
STEP 7 function *Monitor/Modify Variables* for the 1st input data byte and 1st output data byte of the I-slave

6.3 Handling Process Interrupts

Similarly to I/O connected locally in the SIMATIC S7 central rack or expansion rack, distributed I/O devices, too, can generate process interrupts. In a PROFIBUS network, process interrupts can be generated by DP slaves or individual modules contained in a DP slave device, provided the DP slave or I/O module concerned supports interrupt processing. An analog input module with process interrupt capability is thus able to trigger a process interrupt, for instance when a measuring limit value is violated. The user program is interrupted by the process interrupt, and an interrupt OB is called. Note that in SIMATIC S7 a process interrupt is sometimes also referred to as a hardware interrupt.

The following example describes how in a PROFIBUS DP network a slave generates a process interrupt, and how this process interrupt is recognized and evaluated on the DP master. The slave station is an S7-300 programmable controller with CPU 315-2DP as I slave, and the master station is an S7-400 programmable controller.

6.3.1 Generating Process Interrupts on the I Slave (S7-300)

To generate a process interrupt on the related DP master, call system function SFC7 *DP_PRAL* on the CPU 315-2DP station which was configured as the I-slave (see figure 6.10). Note that only the SIMATIC S7 controllers S7-400 and S7-300 with CPU 31x-2DP allow this function.

The module-related input parameters IOID and LADDR of the system function identify the requested process interrupt unambiguously. In our example, we trigger a process interrupt for the output module on the I slave that was given the starting address "1000."

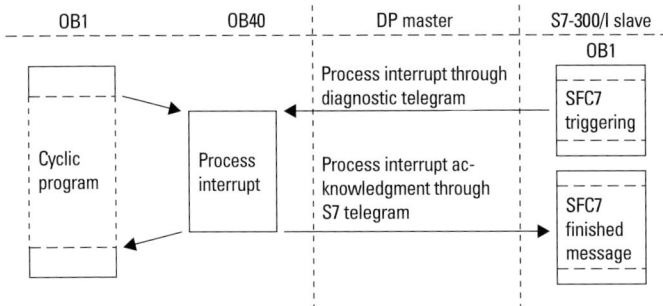

Figure 6.10 Generation of a process interrupt with S7-300 (CPU315-2DP) as I-slave

In our example we are only interested in how the process interrupt is triggered on the I-slave and how it is processed on the DP master. On the I slave we will therefore trigger the process interrupt cyclically. This makes testing and monitoring of the function easier.

We will transmit two pieces of additional information to the DP master. In the first half of its double word, parameter AL_INFO which is an input parameter of SFC7, transmits an application-specific interrupt ID. We will use "ABCD" in our example. And secondly, an interrupt counter (MW106) is transmitted in the parameter's second half of the double word. The counter is incremented with every job that is executed. The interrupt ID is sent to the DP master at the same time as the message of the process interrupt. On the DP master, the interrupt ID is then made available in the local variable OB40_-POINT_ADDR while OB40 is being processed.

To trigger the process interrupt, enter the STL statements shown in figure 6.11 in organization block OB1 for the CPU of the *SIMATIC 300(1)* station. Open the already existing OB1 in the *Blocks* folder and add these statements. Save the block, and quit the editing screen for OB1 in the STEP 7 programming tool *LAD/STL/FBD*.

Then, switch the CPU of the S7-300 station to STOP using the operating mode switch, and download the modified organization block OB1 to the CPU315-2DP. Overwrite the existing one.

```
      L       W#16#ABCD              //Partially preset interrupt identifier
      T       MW104

CALL "DP_PRAL"
      REQ     :=M100.0
      IOID    :=B#16#55              //Address area of module ("55"=output)
      LADDR   :=W#16#3E8             //Start address of module
      AL_INFO:=MD104                 //Application-related interrupt ID
      RET_VAL:=MW102
      BUSY    :=M100.1

      A       M       100.1          //(Cyclic) trigger if SFC7 "free"
      BEC
      =       M       100.0          //Trigger new process interrupt

      L       MW106                  //Increment interrupt counter by one
      +       1
      T       MW106
```

Figure 6.11 Example program on the DP slave S7-300 for generating a process interrupt

6.3.2 Processing Process Interrupts on the DP Master (S7-400)

The process interrupt triggered by the I slave and transmitted by the PROFIBUS DP network is identified by the CPU of the DP master. The operating system of the master CPU calls the related interrupt organization block OB40. The local data of OB40 (see section 5.1.2) contains, among other things, the logical base address of the module that generated the interrupt, and thus provides information about the initiator of the interrupt. With more complex modules, the local data of OB40 also contains information about identifier and status of the interrupt. After execution of the OB40 program (i.e., OB40 concluded), the CPU of the DP master sends an acknowledgment signal to the I-slave which triggered the alarm. This changes the signal status on the BUSY output parameter of system function call SFC7 from "1" to "0".

To evaluate the process interrupt on the DP master, set up organization block OB40 in the *Blocks* folder for the *SIMATIC 400(1)* master station. Type in the STL statements as shown in figure 6.12. Save OB40 and close the editing screen for OB40 in the STEP 7 programming tool *LAD/STL/FBD*.

The load and transfer instructions (see figure 6.12) copy the base address of the interrupting I/O module (submodule) to memory word (MW10), and the user-related interrupt ID to memory double word MD12. Later, you can use the STEP 7 function *Monitor/Modify Variables* to observe interrupt processing by monitoring these two memory areas.

L	#OB40_MDL_ADDR	//Logical base address of the module
T	MW10	
L	#OB40_POINT_ADDR	//Application-specific interrupt ID for //I slave
T	MD12	

Figure 6.12   Example program on the S7-400 DP master for evaluating the process interrupt

Now, download OB40 to the CPU416-2DP of the master station. Then switch the S7-300 CPU to RUN using the operating mode switch (both controllers must now be in RUN mode).

Testing the response of the DP master to a process interrupt

To test the reaction of the DP master to a process interrupt, change in *SIMATIC Manager* to the online view by selecting VIEW → ONLINE. Remember that your PG/PC programming unit must be properly connected to the CPU 416-2DP by the *MPI* cable.

In the *SIMATIC 400(1)* folder, open the *Blocks* folder. Double-click OB40 to obtain the online view of OB40 so that you can observe its execution in STEP 7. In the menu bar, select DEBUG → MONITOR to switch on the status function for OB40 (see figure 6.13). You can now observe how the interrupt is processed on the DP master.

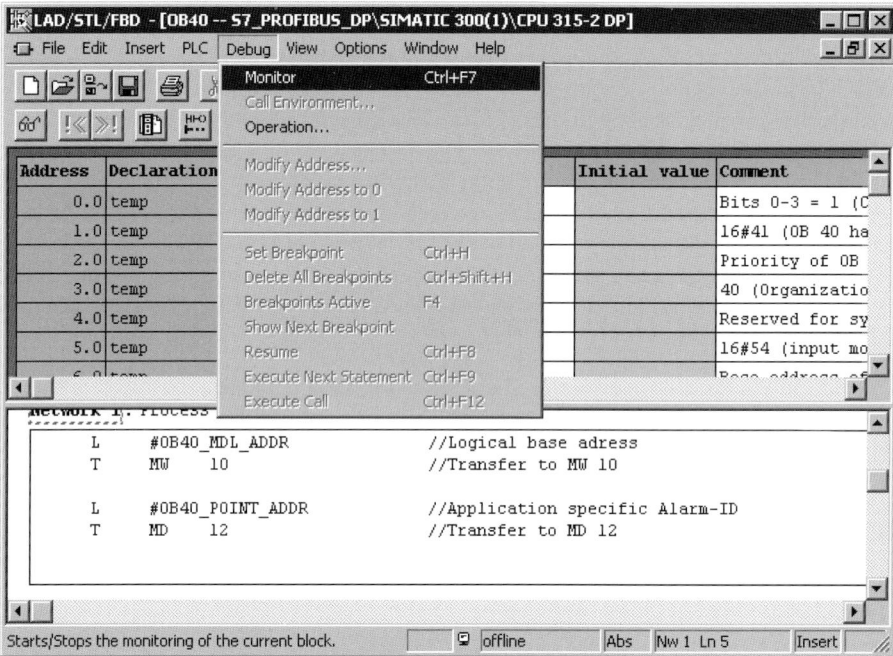

Figure 6.13 Status function for OB40 (example program)

6.4 Transferring Data Records and Parameters

SIMATIC S7 programmable controllers permit the transfer of data records from the user program to SIMATIC S7 modules. This is a method of changing parameter sets of S7 modules during running operation. This online transfer of data records can be applied to both central S7 modules and distributed S7 modules. We distinguish between two types of data records that can be transferred to S7 modules: dynamic data records and static data records. Dynamic data records are usually provided by the user program, whereas static data records are generated in the *HW Config* program and permanently stored in the CPU's system data blocks. SIMATIC S7 offers a number of system functions (SFCs/SFBs) for transferring data records to the S7 modules. See also section 5.5.

The following example describes how to write module data records or parameters to S7 modules. It uses special function blocks SFC55 *WR_PARM* and SFC56 *WR_DPARM* for this purpose. SFC55 transfers a dynamic data record whose contents you can define as required. SFC56 transfers a "static" data record, which is generated with *HW Config* and stored in a system data block (SDB) on the CPU. During system startup, this data record is transferred automatically to the appropriate module.

In our example we will use system function SFC55 to change the measuring range of the analog input module on the ET200M station which we configured earlier in section 4.2.5. The measuring range shall be changed from +/–10 V to +/–2.5 V. Afterward, we will cancel the change in the parameter set by calling SFC56 so that the module will again use the parameters specified before, during configuration in *HW Config*.

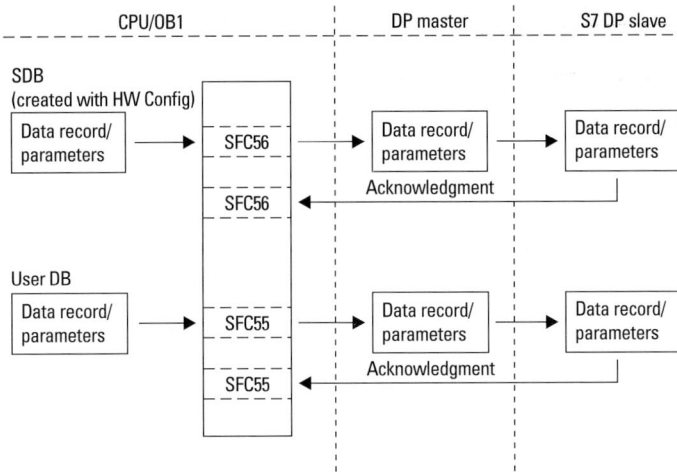

Figure 6.14   Transferring a data record to an S7 module by calling SFC55/SFC56

In practice, this function could be used for instance when a measured input approaches a certain state or value and a higher resolution is required for a short period.

6.4.1 Layout of the Data Record (DR1) for the Analog Input Modules of SIMATIC S7-300

The analog input module used in our example is the module "SM331 AI2×12 Bit" of the SIMATIC S7-300 series. It has two analog input channels with a resolution of 12 to 14 bits. Table 6.1 shows the data records of the analog input modules for the SIMATIC S7-300 controllers. Since data record no. "0" (DR0) can only be read by system functions, but not written, it cannot be transferred to the analog input modules by means of SFC55.

Table 6.1   Data records and parameters of the analog input modules of the SIMATIC S7-300 series

Parameter	Data Record No.	Parameters Can Be Set with SFC55
Group diagnosis	0	No
Diagnosis incl. wire-break check	0	No
Limit value alarm enable		Yes
Diagnosis alarm enable		Yes
Interference frequency suppression		Yes
Measuring type	1	Yes
Measuring range		Yes
Upper limit		Yes
Lower limit		Yes

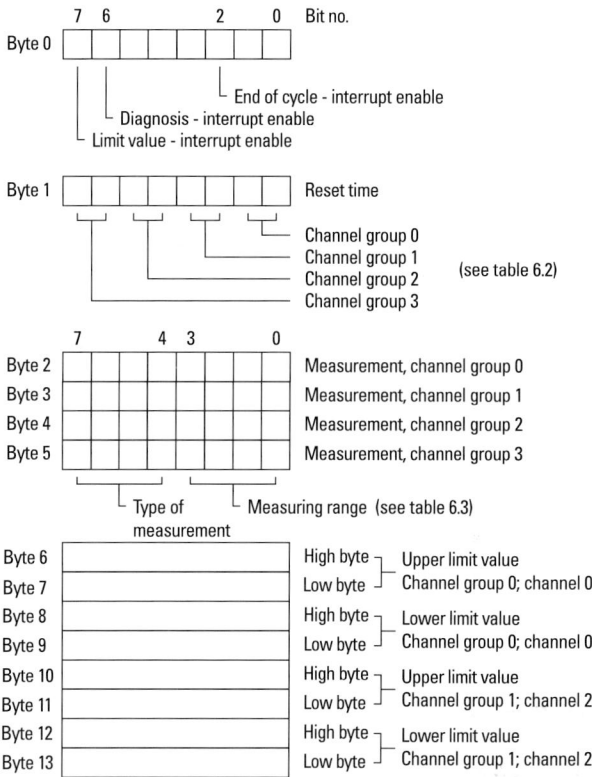

Figure 6.15 Layout of data record DR1 for the analog input modules of the S7-300

Figure 6.15 illustrates in detail how data record DR1 is structured when it contains parameters for the analog input modules of the SIMATIC S7-300 controller. The parameters stored in DR1 can be used to enable alarms and interrupts, select the integration time for the suppression of interference frequencies, set the type and range of measurement, and set the upper and lower limit values of the measuring range for the analog input channel groups, if applicable. DR1 has a length of 14 bytes.

Table 6.2 shows the possible settings of the integration time for interference frequency suppression on the analog input modules.

Table 6.2 Integration times which can be set for analog input modules of the S7-300 series

Interference Frequency Suppression	Integration Time	Setting
400 Hz	2.5 msec	2#00
60 Hz	16.7 msec	2#01
50 Hz	20.0 msec	2#10
10 Hz	100.0 msec	2#11

Table 6.3 shows the measuring ranges which can be set for measuring type "Voltage" of the analog input modules for S7-300 controllers.

Table 6.3   Measuring ranges for measuring type "Voltage" for S7-300 analog input modules

Type of Measurement	Setting	Measuring Range	Setting
Voltage	2#0001	± 80 mV	2#0001
		± 250 mV	2#0010
		± 500 mV	2#0011
		± 1 V	2#0100
		± 2.5 V	2#0101
		± 5 V	2#0110
		1 to 5 V	2#0111
		± 10 V	2#1001
		± 25 mV	2#1010
		± 50 mV	2#1011

In our example project configured in *HW Config* we have set the following values for the analog input modules on the ET200M station.

Diagnosis	Group diagnosis "ON"
Type of measurement	Voltage (V)
Measuring range	+/–10 V
Integration time	20 msec

6.4.2 Example: Changing the Parameters of Analog Input Modules Using SFC55 *WR_PARM*

The following example for the use of system function SFC55 is based on our previous example project "ET200M" described in section 4.2.5. However, since we now only use the S7 DP master station S7-400 and the DP slave station ET200M, you will have to delete the ET200B station and the S7-300 station configured earlier in *HW Config*. Connect the DP interfaces of the S7-400 controller and the ET200M station using a PROFIBUS cable, and switch on the power supplies of the devices. Our example is based on the assumption that an overall reset has been carried out on the DP master controller and this controller is in RUN mode (i.e., operating mode switch in position RUN-P). In addition, we assume that the PROFIBUS address on the ET200M station has been set to "5."

In the *SIMATIC 400(1)* folder, open the *Blocks* folder and set up data block DB30 with the structure shown in table 6.4. Save the block, and quit the editing screen for this block.

Table 6.4
Data record 1 for the analog input module for changing the measuring range to +/- 2,5 V

Byte No.	Name	Type	Initial Value	Comment
0.0		STRUCT		
+ 0.0	AlarmEnable	BYTE	B#16#00	Limit value/diagnostic interrupt
+ 1.0	IntTime	BYTE	B#16#02	Integration time: 20 msec
+ 2.0	M_Kgr_0	BYTE	B#16#15	Channel group 0 (voltage; +/- 2,5 V)
+ 3.0	M_Kgr_1	BYTE		Channel group 1 (not relevant)
+ 4.0	M_Kgr_2	BYTE		Channel group 2 (not relevant)
+ 5.0	M_Kgr_3	BYTE		Channel group 3 (not relevant)
+ 6.0	OGr_Kgr_0H	BYTE		
+ 7.0	OGr_Kgr_0L	BYTE		Limit values not relevant since these were not enabled.
+ 8.0	UGr_Kgr_0H	BYTE		
+ 9.0	UGr_Kgr_0L	BYTE		
+ 10.0	OGr_Kgr_1H	BYTE		Does not exist
+ 11.0	OGr_Kgr_1L	BYTE		Does not exist
+ 12.0	UGr_Kgr_1H	BYTE		Does not exist
+ 13.0	UGr_Kgr_1L	BYTE		Does not exist
= 14.0		END_-STRUCT		

In *SIMATIC Manager*, change to the offline view by selecting *View → Offline* in the menu bar. In the *SIMATIC 400(1)* folder, open the *Blocks* folder and OB1. Enter the system function SFC55 *WR_PARM* as shown in figure 6.16. Save OB1 and close the editing screen in *LAD/STL/FBD*.

Then change back to *SIMATIC Manager*. Use the DOWNLOAD button from the toolbar to copy all blocks contained in the *SIMATIC 400(1)* folder to the CPU416-2DP. Remember that your PG programming unit or PC must be connected to the CPU416-2DP with an *MPI* cable.

```
CALL "WR_PARM"
    REQ     := M30.0              //Trigger job
    IOID    := B#16#54            //Identifier for I/O input module
    ADDR    := W#16#200           //Address of input module (512dec)
    RECNUM  := B#16#1             //Data record number (DR1)
    RECORD  := P#DB30.DBX 0.0 BYTE 14   //Pointer to DR1 in DB 30
    RET_VAL := MW32
    BUSY    := M30.1

    A       M30.1                 //Reset job trigger
    R       M30.0
```

Figure 6.16 Call of SFC55 to change the parameter set for the analog input module

After download, the CPU416-2DP must be in the RUN mode, and none of the DP-related error LEDs ("SF DP" LED or "BUSF" LED) should be on or flashing. This also applies to the LEDs of the ET200M station. If all DP-related error LEDs are off, user data communication between the DP master and the ET200M station is running correctly.

6.4.3 Testing the Parameters for the Analog Input Module Changed with SFC55 *WR_PARM*

Use the STEP 7 function *Monitor/Modify Variables* (see section 6.2.3) to call the programmed system function SFC55 and observe how this SFC changes the measuring range of the analog input module on the ET200M station from +/–10 V to +/–2.5 V.

In the variable table under "Address", enter the two variables MB30 (M30.0=REQ and M30.1=BUSY) and MW32 (RET_VAL). Specify a modify value of B#16#01 for MB30. Activate the display of the monitor value by selecting VARIABLE → DISPLAY FORCE VALUES in the menu bar. The monitor value for MB30 is B#16#00, whereas the monitor value for RET_VAL (MW32) must indicate the value W#16#7000. Select ACTIVATE MODIFY VALUES to activate the value entered for MB30. This starts the programmed system function SFC55.

Immediately after the force procedure, the two variables must again contain the initial monitor values. This indicates that the function was executed correctly.

Remarks: When the DP master system is restarted, the parameter changes for the analog input module carried out in this way are lost. In this case, the analog input module will receive its parameter set from the static DR1 stored in the system data block.

6.4.4 User Program for Changing the Parameter Set for the Analog Input Module Using SFC56 *WR_DPARM*

SFC56 *WR_DPARM* transfers the original module parameters defined in *HW Config* to the analog input module on the ET200M station. This parameter set is kept in data record DR1. DR1 is predefined for the analog input module and stored in the applicable SDB on the CPU.

To program the system function call, open the *SIMATIC 400(1)* folder, followed by the *Blocks* folder, and open organization block OB1. Enter the call of SFC56 *WR_DPARM* in STL format as shown in figure 6.17. Save OB1 and close the editing screen in the STEP 7 tool *LAD/STL/FBD*.

```
CALL "WR_DPARM"
    REQ      : =M40.0        //Job trigger
    IOID     : =B#16#54      //Identifier for I/O input module
    LADDR    : =W#16#200     //Address of input module (512dec)
    RECNUM   : =B#16#1       //Data record number (DR1)
    RET_VAL  : =MW42
    BUSY     : =M40.1

    A          M40.1         //Reset job trigger
    R          M40.0
    R          M40.0
```

Bild 6.17   Call of SFC56 *WR_DPARM* in OB1

Return to *SIMATIC Manager*, and use DOWNLOAD to copy all blocks from the *Blocks* folder of *SIMATIC 400(1)* to the CPU416-2DP. Make sure that your PG programming unit or PC is properly connected to the CPU416-2DP by means of an *MPI* cable.

After download, the CPU416-2DP must be in RUN mode, and none of the DP-related error LEDs ("SF DP" LED or "BUSF" LED) should be on or flashing. The same applies to the LEDs on the ET200M station. If all DP-related error LEDs are off, user data communication between the DP master and the ET200M station is running correctly.

6.4.5 Testing the Parameters for the Analog Input Module Changed with SFC56 *WR_DPARM*

Use the STEP 7 function *Monitor/Modify Variables* to call the programmed SFC56 and observe how the SFC56 restores the parameter set of the analog input module on the ET200M station to its original state.

In the variable table, enter the two variables MB40 (M40.0=REQ and M40.1=BUSY) and MW42 (RET_VAL). Specify a monitor value of B#16#01 for MB40. Select DISPLAY FORCE VALUES to display the monitor value. The monitor value for MB40 shows B#16#00. The status value for RET_VAL (MW42) must show the value W#16#7000. Select ACTIVATE MOFIFY VALUES to activate the value entered for MB40. This starts the programmed SFC56. Immediately after the force procedure, the two variables again contain the entered force values. This indicates that the function has been executed correctly.

6.5 Triggering the DP Control Commands SYNC/FREEZE

The control commands SYNC (synchronizing the outputs) and FREEZE (freezing the inputs) allow you to coordinate data communication with several slaves.

A DP master with appropriate functionality can simultaneously send the control commands (Broadcast telegrams) SYNC and/or FREEZE to a group of DP slaves. The DP slaves are combined for this purpose into SYNC and FREEZE groups. Up to 8 groups can be created for one master system. However, each DP slave can only be assigned to one group.

Use the SYNC control command if you want to synchronize the outputs on several slaves. On receiving the SYNC control command, the addressed DP slaves read the data which they have received with the last Data_Exchange telegram from the DP master and which they have stored in their transfer buffers. They then switch this data to the outputs. This permits simultaneous activation (synchronization) of the output data on several DP slaves. Figure 6.18 shows the principal sequence of a SYNC command.

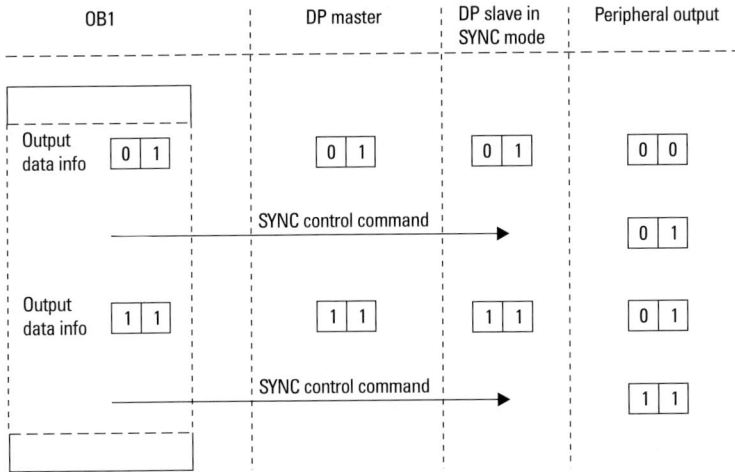

Figure 6.18   Principal sequence of a SYNC command

The UNSYNC control command cancels the SYNC mode for the addressed DP slaves. The DP slave returns to cyclic data transfer. This means that the data sent by the DP master is immediately switched to the outputs.

Use the FREEZE control command if you want to freeze the input data of DP slaves. When a FREEZE command is sent to a group of DP slaves, all these DP slaves simultaneously freeze the signals currently present on their inputs so that these can be read by the DP master. The input data on the DP slaves is not updated until another FREEZE command is received. Figure 6.19 shows the sequence of a FREEZE command.

The UNFREEZE control command cancels the FREEZE mode for the addressed DP slaves so that they return to cyclic data transfer with the DP master. The input data is immediately updated by the DP slave and can then be read immediately by the DP master.

Remember also that, after a hot or warm restart, a DP slave does not switch to SYNC or FREEZE mode until the first SYNC or FREEZE command sent by the DP master is received.

OB1	DP master	DP slave in FREEZE mode	Peripheral input

Input data info `0` `0`   `0` `0`   `0` `0`   `0` `1`

FREEZE control command →   `0` `1`

Input data info `0` `1`   `0` `1`   `0` `1`   `0` `1`

Input data info `0` `1`   `0` `1`   `0` `1`   `1` `1`

FREEZE control command →   `1` `1`

Input data info `1` `1`   `1` `1`   `1` `1`   `1` `1`

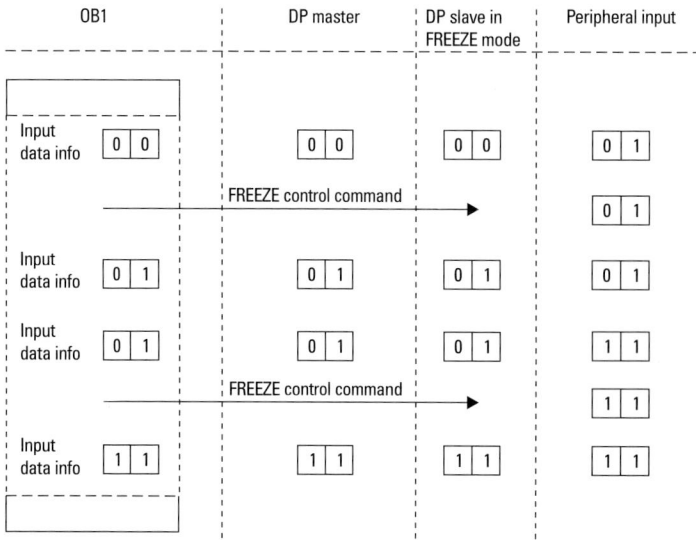

Figure 6.19 Principal sequence of a FREEZE command

6.5.1 Example of SYNC/FREEZE with DP Master IM 467

The following example illustrates the use of the **SYNC/FREEZE** control commands.

To create the required plant configuration, open *SIMATIC Manager*, and in the menu bar select FILE → NEW. Give the new project the name "SYNCFR", and quit the screen with OK. Using INSERT→ STATION → SIMATIC 400-STATION, enter a new S7-400 station.

Double-click the *SIMATIC 400(1)* folder to open the station. The *Hardware* object appears in the right-hand screen of *SIMATIC Manager*. Double-click this object to open the hardware configuration of the SIMATIC 400 station.

Insert rack "UR2" from the hardware catalog. Place power supply "PS407 10A" in slot 1. When selecting the CPU, remember that this device must support the SYNC and FREEZE functions. For example, select CPU 416-1 with the order number 6ES7416-1XJ02-0AB0, and place it in slot 3.

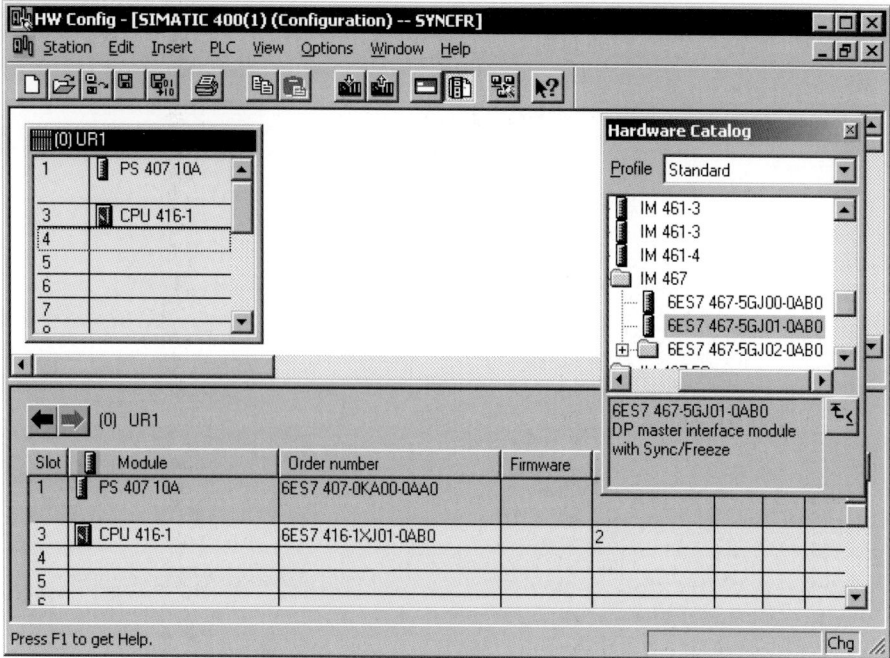

Figure 6.20   Selecting IM 467 from the hardware catalog

To configure the plug-in DP master module (IM 467), go to the hardware catalog of the SIMATIC 400 station and open the sub-catalog IM-400. Select module IM 467 with order number 6ES7467-5GJ01-0AB0, and place it in slot 4. See figure 6.20.

When the modules are placed in the subrack, the dialog box "Properties PROFIBUS Station IM 467" and the "Network connection" tab automatically appear on the screen. Select NEW, and confirm the dialog box with OK. This creates a new PROFIBUS subnet with a transmission rate of 1.5 Mbaud and a bus parameter profile of type "DP." Confirm the suggested station address "2" for the IM 467 and close the dialog box with OK. The IM 467 module is now inserted in slot 4, and the DP master system for the IM 467 is shown graphically. See figure 6.21.

In the next step we will configure the slaves. We will use simple ET 200B stations which support the SYNC and FREEZE control commands. In the Hardware folder, open the selection of PROFIBUS DP modules. In the sub-catalog "ET 200B" select module "B-16D1."

Drag the module to the DP master system of the IM 467 displayed on the screen. The "Properties of PROFIBUS station ET 200B 16D1" dialog box is opened. Select "3" as the PROFIBUS address, and quit the screen with OK.

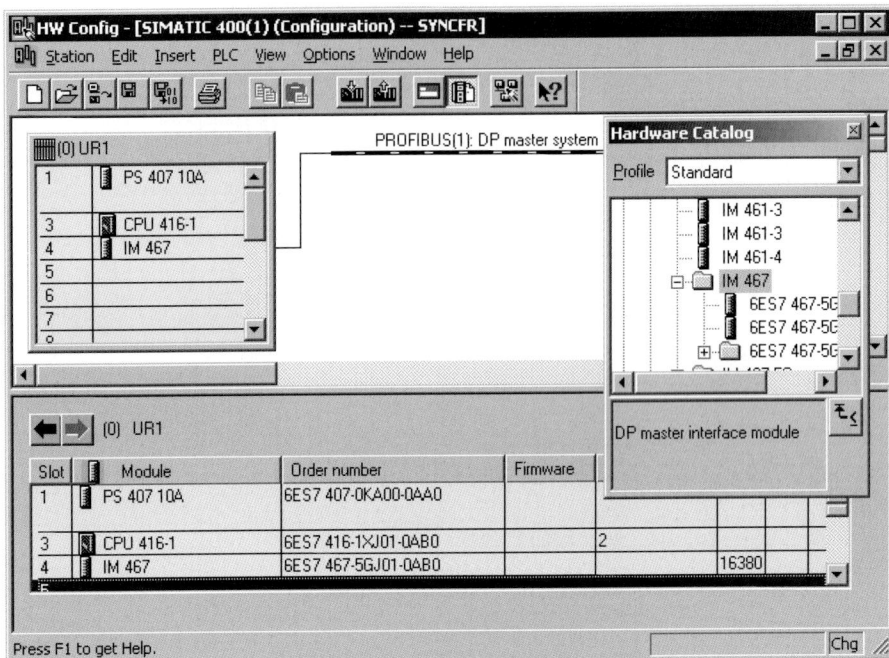

Figure 6.21   Hardware configuration with IM 467

Drag the "B-16DO" module from the hardware catalog PROFIBUS DP → ET 200B to the master system of the IM 467. In the "Properties of PROFIBUS station ET200B 16DO" dialog box, set the PROFIBUS address to "4", and close the screen with OK.

The configuration of the DP master system of the IM 467 for our example project is now complete.

Next, we will define the settings for the SYNC/FREEZE function. To do this, double click the DP master system PROFIBUS(1) displayed on the screen.

The "DP Master System Properties" dialog box and the "Group Assignment" tab appear. Here, you can assign the DP slaves with SYNC/FREEZE capability to various groups. See figure 6.22. The first column of the table contains the DP slaves configured on the DP master system. They are arranged in the order of their PROFIBUS address (PROFI-BUS address in parentheses). Columns 1 to 8 contain 8 possible groups to which you can assign the DP slaves.

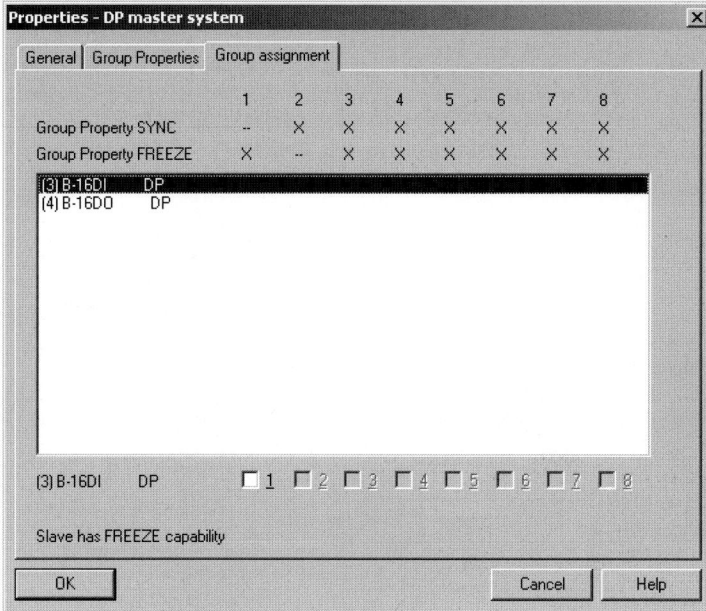

Figure 6.22   Group assignment in *HW Config*

First select the "Group Assignment" tab to specify the characteristics of the groups used. In the "Comment" column, you can specify an extra text (comment/group designation) for the particular group. In the "Properties" column, select the function you want to assign to the group. Define the parameters of the groups as shown in figure 6.23. Group 1 is defined as the FREEZE group, and group 2 is defined as the SYNC group.

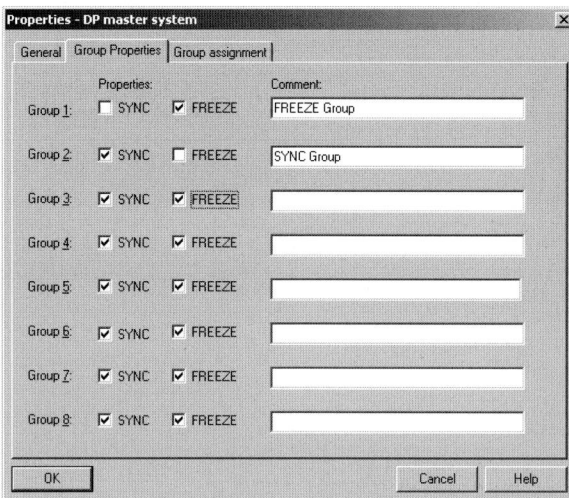

Figure 6.23   Group properties in *HW Config*

163

Click on the "Group Assignment" tab to accept the changes and go to the Properties window of Group Assignment. Click station B-16DI. You can now assign the DP slave to group 1. Then select the B16-DO DP slave and assign it to group 2 (see figure 6.24). Confirm the settings with OK.

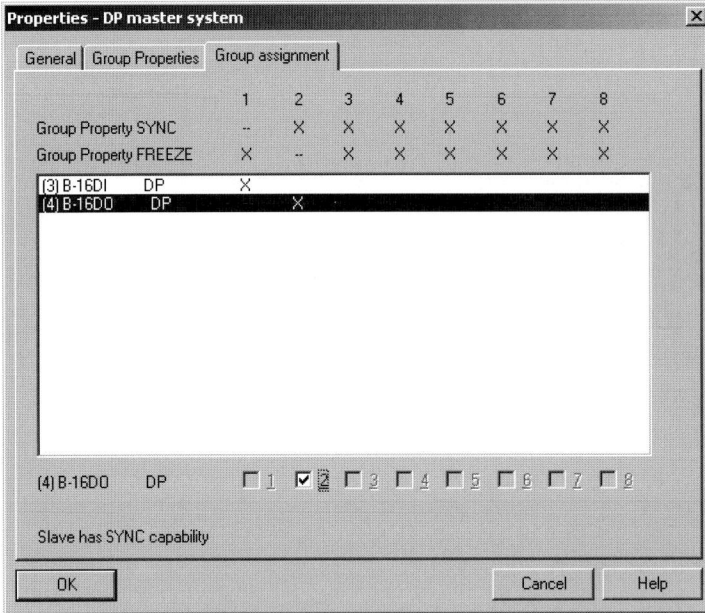

Figure 6.24 Configured group assignment with ET 200B modules

Configuration of the DP master system is now complete.

Select STATION → SAVE AND COMPILE. Switch your SIMATIC 400 station to STOP, and download the hardware configuration to the S7-400 CPU. The hardware setup of your SIMATIC 400 station must of course match the project which you have configured in *HW Config*.

Connect the IM467 module to the two ET 200B modules using a PROFIBUS cable, and change the operating mode switch of the CPU416-1 to RUN-P. The CPU goes to RUN. All red error LEDs must be off. Close the *HW Config* program.

6.5.2 Generating the User Program for the SYNC/FREEZE Function

In the final step we will program the system function SFC11 for the SYNC/FREEZE function. In our example, we will call SFC11 *DPSYC_FR* in OB1 on a signal edge change. Back in *SIMATIC Manager*, double-click the CPU416-1 now shown in the right-hand window. The object is opened, and the "S7 program(1)" folder appears. Double-click "S7 program(1)" and open the "Blocks" folder. The default OB1 appears (see figure 6.25).

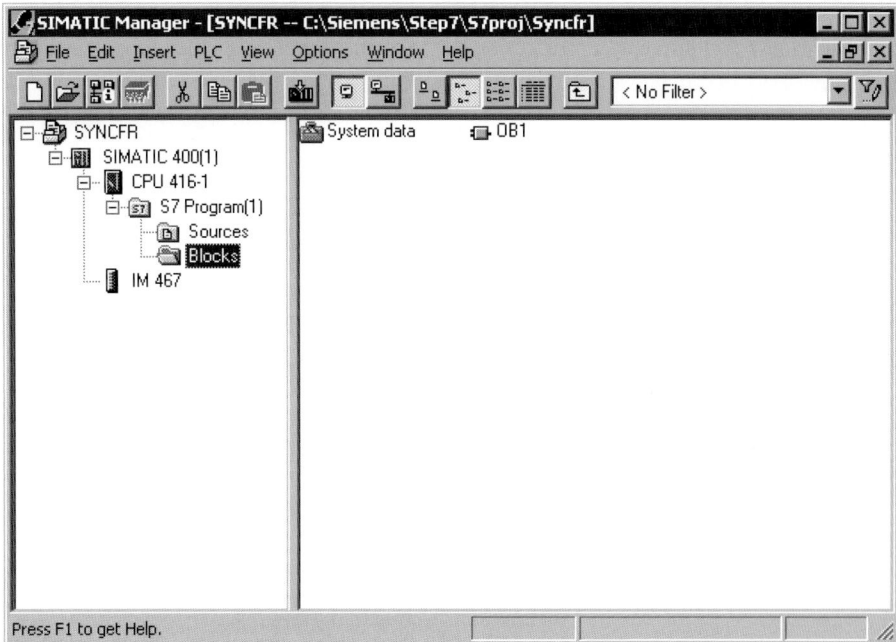

Figure 6.25 *SIMATIC Manager* with *Blocks* folder open

Open OB1 with a double-click. The "Properties – Organization Block" dialog box for OB1 appears. Click OK to start the STEP 7 program *LAD/STL/FBD* for programming OB1 in the STL view.

To be able to use SFC11 from the standard library V3.x, select VIEW→ CATALOG. In the list of "Program Elements", select LIBRARIES → STANDARD LIBRARY → SYSTEM FUNCTION BLOCKS (see figure 6.26).

Select SFC11 *DPSYC_FR* and drag it to the first network of OB1. Complete the STL program as shown in figure 6.27.

Save and download OB1 to CPU416-1. You can now monitor the program using the STEP7 function *Monitor/Modify Variables*. For this, select PLC →MONITOR/MODIFY VARIABLES in the screen *LAD/STL/FBD*.

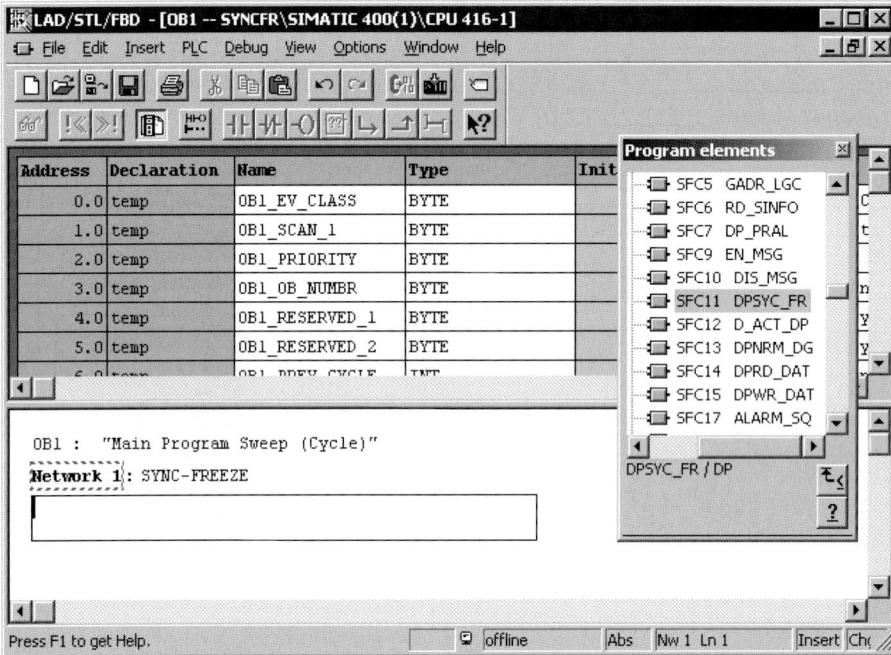

Figure 6.26 *Program LAD/STL/FBD* – List of system function blocks

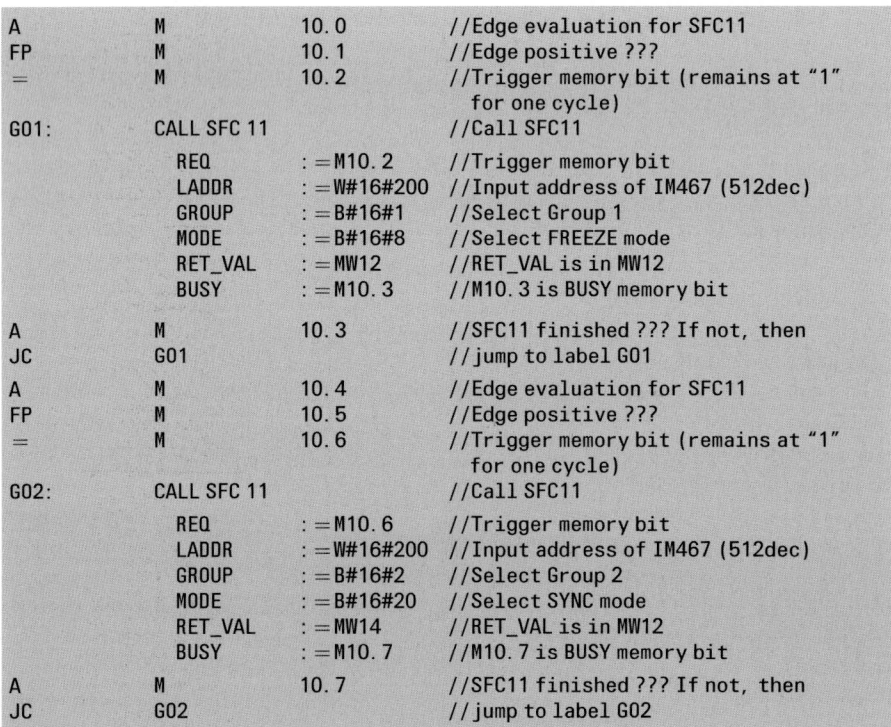

A	M	10. 0	//Edge evaluation for SFC11
FP	M	10. 1	//Edge positive ???
=	M	10. 2	//Trigger memory bit (remains at "1" for one cycle)
G01:	CALL SFC 11		//Call SFC11
	REQ	: =M10. 2	//Trigger memory bit
	LADDR	: =W#16#200	//Input address of IM467 (512dec)
	GROUP	: =B#16#1	//Select Group 1
	MODE	: =B#16#8	//Select FREEZE mode
	RET_VAL	: =MW12	//RET_VAL is in MW12
	BUSY	: =M10. 3	//M10. 3 is BUSY memory bit
A	M	10. 3	//SFC11 finished ??? If not, then
JC	G01		// jump to label G01
A	M	10. 4	//Edge evaluation for SFC11
FP	M	10. 5	//Edge positive ???
=	M	10. 6	//Trigger memory bit (remains at "1" for one cycle)
G02:	CALL SFC 11		//Call SFC11
	REQ	: =M10. 6	//Trigger memory bit
	LADDR	: =W#16#200	//Input address of IM467 (512dec)
	GROUP	: =B#16#2	//Select Group 2
	MODE	: =B#16#20	//Select SYNC mode
	RET_VAL	: =MW14	//RET_VAL is in MW12
	BUSY	: =M10. 7	//M10. 7 is BUSY memory bit
A	M	10. 7	//SFC11 finished ??? If not, then
JC	G02		// jump to label G02

Figure 6.27 Listing of OB1 with SFC11 *DPSYC_FR*

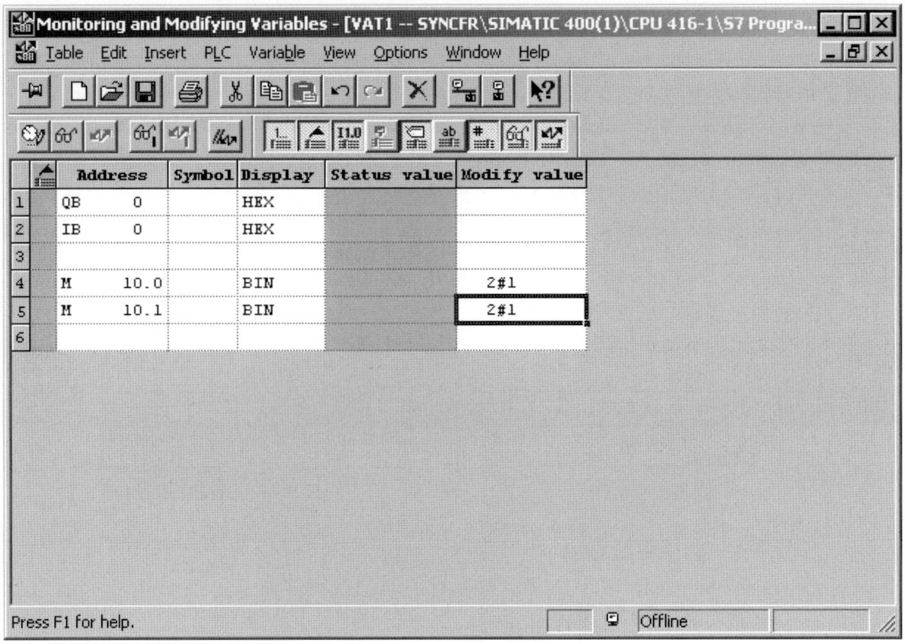

Figure 6.28 Variable table for testing SFC11 *DPSYC_FR*

Type in the lines as shown in figure 6.28. "QB0" is the first output byte of the ET 200B/16DO module, and "IB0" is the first input byte of the ET 200B/16DI module. Memory bit M 10.0 has the task of triggering the job for the FREEZE group, and memory bit M 10.4 triggers the job for the SYNC group.

After startup of the DP bus system, all DP slaves transfer data cyclically. Setting memory bits M 10.1 and M 10.4 to signal status "1" invokes the SYNC and FREEZE control commands.

The ET 200B/16DI module is now in FREEZE mode, and the ET 200B/16DO module is in SYNC mode. Changes in the input signals on the ET 200B/16DI station are now no longer passed on to the CPU, and you cannot observe these changes in the *Monitor/Modify Variables* dialog box.

Similarly, the values which are entered for "QB0" and activated with ACTIVATE FORCE VALUES are not switched to the outputs of the ET 200B/16DO module. The SYNC and FREEZE control commands are not triggered again until the job trigger memory bits M 10.0 and M 10.4 change from signal status "0" back to signal status "1" during the SFC11 call. This transfers to the outputs the output data that has been set and transferred to the ET 200B/16DO station, and reads in the current input data of the ET 200B/16DI module.

Remember, however, that the outputs of the DP slaves which were addressed with system function SFC11 may not be changed by the user program while an SFC11 is running (BUSY="1"). We recommend that you program SFC11 in a loop (scan BUSY) or use the "partial process image" function.

167

6.6 Exchanging Data Using Cross Communication

Cross communication is used to forward input data of a DP slave directly to other DP slaves and a Class 2 DP master. With cross communication, the DP slave sends its response telegram to the DP master through a *one-to-many* connection instead of a *one-to-one* connection (see figure 6.29).

Again, use the *HW Config* program to configure the cross communication connections. Note that you can only use those DP bus stations (master/slave) for cross communication that support this function.

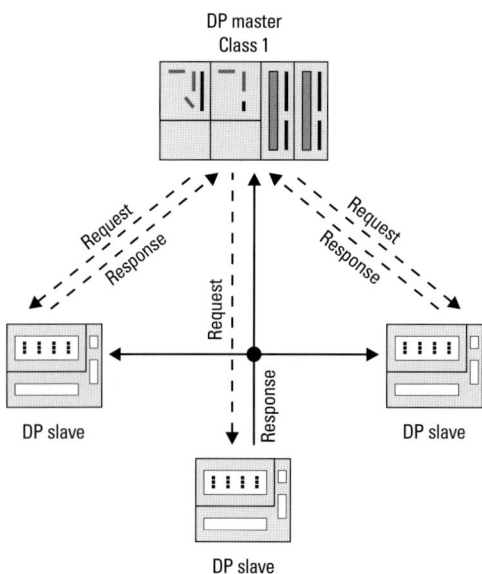

Figure 6.29   Response telegram of the DP slave in cross communication mode

6.6.1 Example Project for Cross Communication with I Slaves (CPU 315-2DP)

The following example describes the use of cross communication connections. It illustrates slave-to-slave and slave-to-DP-master data communication using the S7-300-CPU315-2DP as DP master and I slaves.

To create the required hardware configuration, open *SIMATIC Manager*, and select *File → New*. Enter the name "Cross Communication" for the new project, and quit the dialog box with OK. In the menu bar, select *Insert → Station → SIMATIC 300 Station* to insert a new S7-300 station. Give it the name "DP Master." Using the same procedure, add three more S7-300 stations with the names "I-slave 5," "I-slave 6" and "DP Master/Inputs" (see figure 6.30).

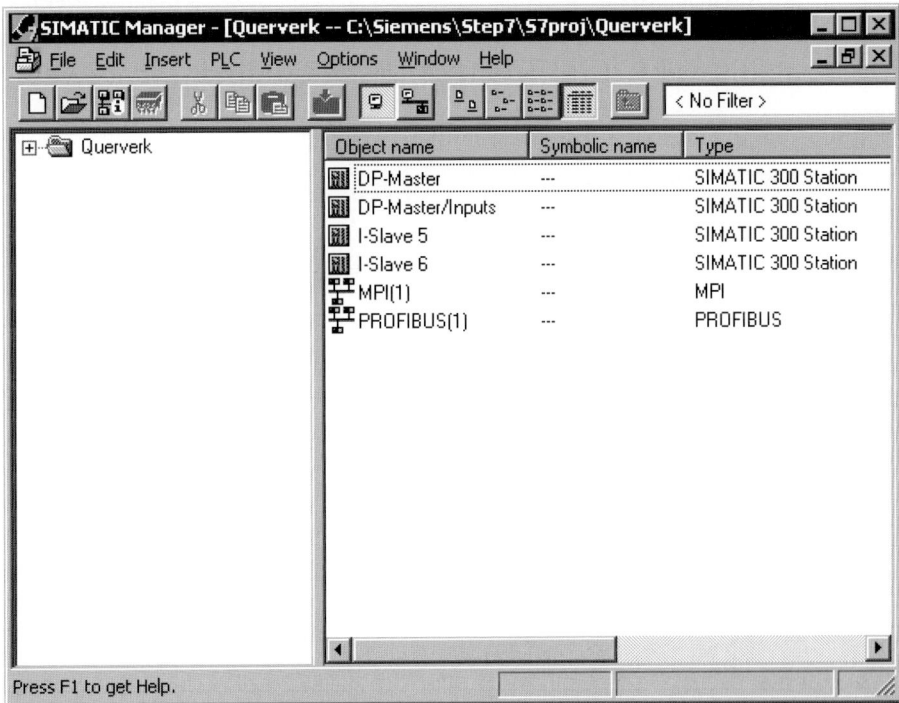

```
SIMATIC Manager - [Querverk -- C:\Siemens\Step7\S7proj\Querverk]    _□X
File  Edit  Insert  PLC  View  Options  Window  Help                _ɒ X

  □ ᴄ̃ 𝔐 ▦   ▒ ▣ ▒   ▦ ▣ ▣ᴏ   ᴅ ᴅ ᴏ ᴅ   ▦   ▦   < No Filter >

 ⊞  Querverk            │ Object name      │ Symbolic name │ Type
                        │ ▦ DP-Master      │ ---           │ SIMATIC 300 Station
                        │ ▦ DP-Master/Inputs│ ---          │ SIMATIC 300 Station
                        │ ▦ I-Slave 5      │ ---           │ SIMATIC 300 Station
                        │ ▦ I-Slave 6      │ ---           │ SIMATIC 300 Station
                        │ ⊞ MPI(1)         │ ---           │ MPI
                        │ ⊞ PROFIBUS(1)    │ ---           │ PROFIBUS

 Press F1 to get Help.
```

Figure 6.30 Project "Cross Communication" (Querverkehr) with SIMATIC 300 stations

Double-click the *I-slave 5* folder to open the first S7-300 DP slave station. The Hardware object appears in the right-hand window of *SIMATIC Manager*. Double-click this folder to open the hardware list of the SIMATIC 300 station. In this list, select RACK-300 and then select and move the component "Rail" to the upper section of the station screen. Next, move the "PS 307 2A" load current power supply to slot "1" of the rack. When selecting a CPU, remember that this device must be able to support cross communication. Therefore, select CPU 315-2DP with order number 6ES7 315-2AF03-0AB0, and drag it to slot "2" of the rack.

When the CPU is inserted in slot 2, the "Properties PROFIBUS Node DP Master" is automatically opened on the screen. On the "Parameters" tab, change the preset PROFIBUS address to "5." On the right-hand side of the "Subnet" group, press the "New ..." button. The "Properties – New Subnet PROFIBUS" dialog box appears. Confirm the "General" tab with OK. Next, confirm the "Parameters" tab with OK. This creates a new PROFIBUS subnet with a rate of transmission of 1.5 Mbps and a bus parameter profile of type "DP."

Now, double-click the DP master interface of the CPU 315-2DP to call the "Properties – DP Master" dialog box. In the "Operating Mode" tab, set the DP interface of the CPU to "DP Slave."

Now change to the "Configuration" tab. Via the respective configuration dialog, enter all data communication settings required for the I-slave. In the "Mode" field of the configuration dialog, define whether the I/O data specified in the next column is to be exchanged through a master-slave connection (MS = Master-Slave) or through a cross connection (DX = Direct Communication). Enter the parameters and values shown in figure 6.31 into the configuration dialog, and quit the screen with OK. Save the *HW Config* parameter set for this slave using *Station → Save and Compile* from the menu bar.

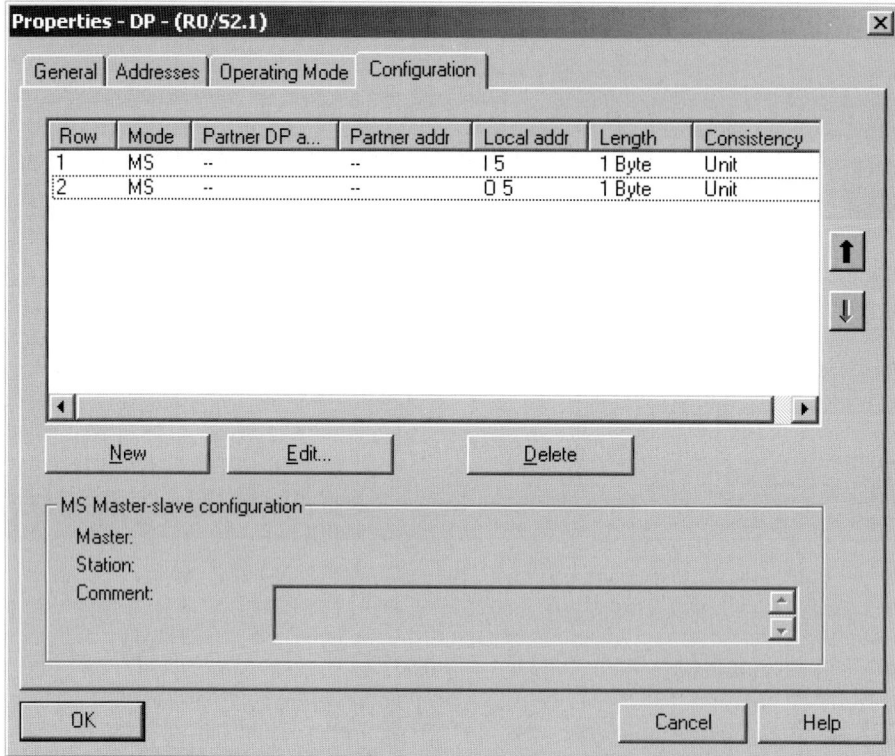

Figure 6.31   Configuration of I-slave 5

Now, back in *SIMATIC Manager*, configure I-slave 6 in the same way. Set the PROFI-BUS address to "6," and add the slave to the already existing PROFIBUS subnet "PROFIBUS(1)". On the "Configuration" tab, set the values as shown in figure 6.32. Save and compile the configuration for I-slave 6.

Configure the hardware for the S7-300 DP master station in the same way. Enter PROFI-BUS address "2" for this station, and connect the master to the already existing PROFI-BUS subnet "PROFIBUS(1)." Since this station is a DP master, keep the preset "DP Master" mode on the "Operating Mode" tab.

In the next few steps, you will connect the two DP slave stations just configured (I-slave 5 and I-slave 6) to the PROFIBUS DP subnet of the DP master.

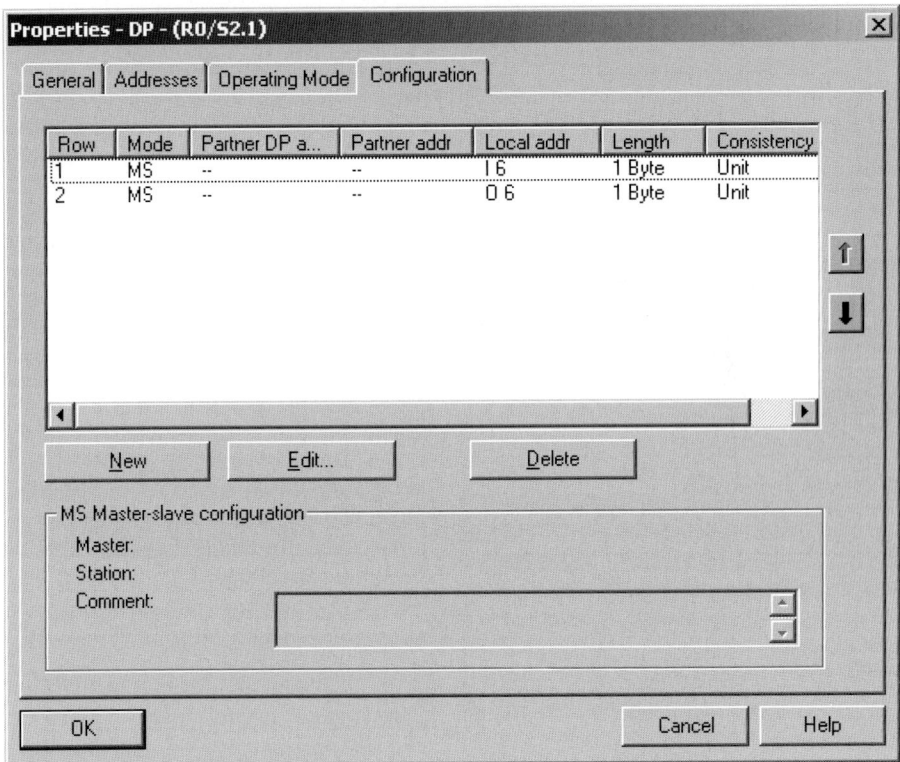

Figure 6.32 Configuration of I-slave 6

In the hardware catalog under "PROFIBUS DP", open the sub-catalog "Configured Stations." Select the CPU31x-2DP and connect it to the DP master system using a drag-and-drop operation. In the "DP Slave Properties" dialog box (figure 6.33), on the "Connection" tab, select station "I-slave 5" and connect it to the DP master system using the "Connect" button.

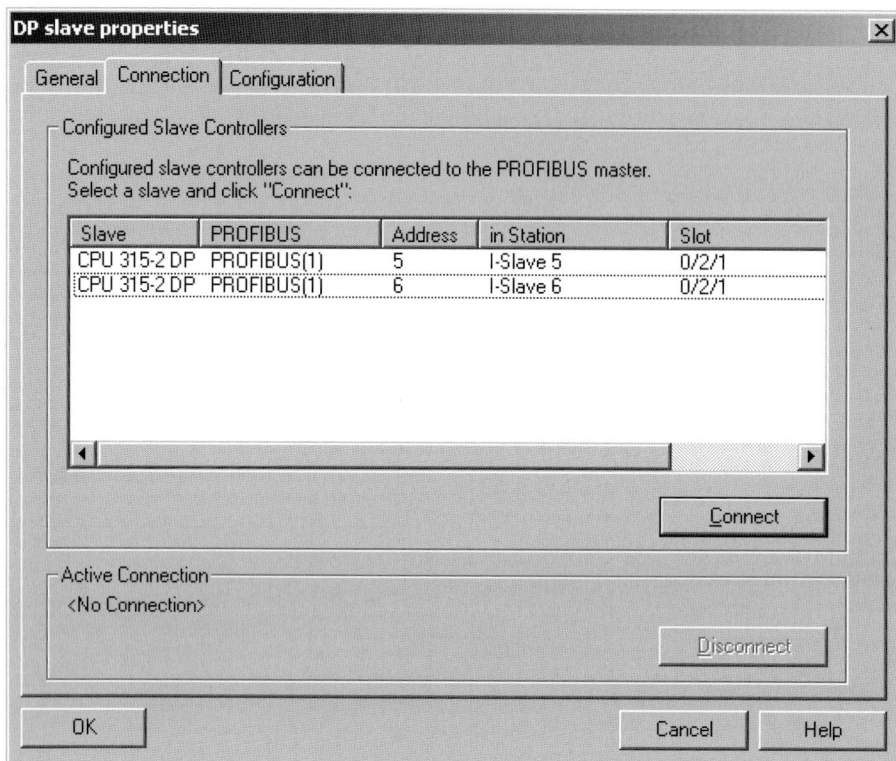

Figure 6.33   Connecting I-slave 5 to the PROFIBUS subnet

On the "Configuration" tab, complete the I/O configuration for I-slave 5 in the section "PROFIBUS DP Partner" (figure 6.34). These are the I/O characteristics as seen from the DP master. Close the "DP Slave Properties" dialog box with OK.

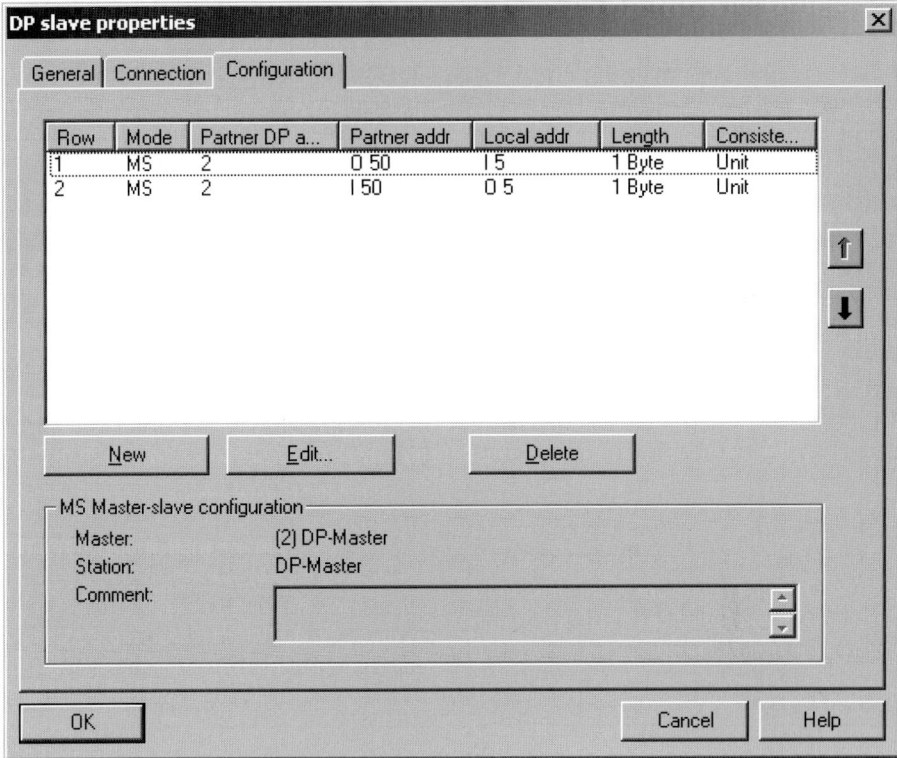

DP slave properties ✕

General | Connection | Configuration

Row	Mode	Partner DP a...	Partner addr	Local addr	Length	Consiste...
1	MS	2	O 50	I 5	1 Byte	Unit
2	MS	2	I 50	O 5	1 Byte	Unit

New   Edit...   Delete

MS Master-slave configuration
Master: (2) DP-Master
Station: DP-Master
Comment:

OK   Cancel   Help

Figure 6.34   I/O configuration of I-slave 5

173

Using the same procedure, connect station I-slave 6 to the DP master system, and complete the I/O configuration as shown in figure 6.35.

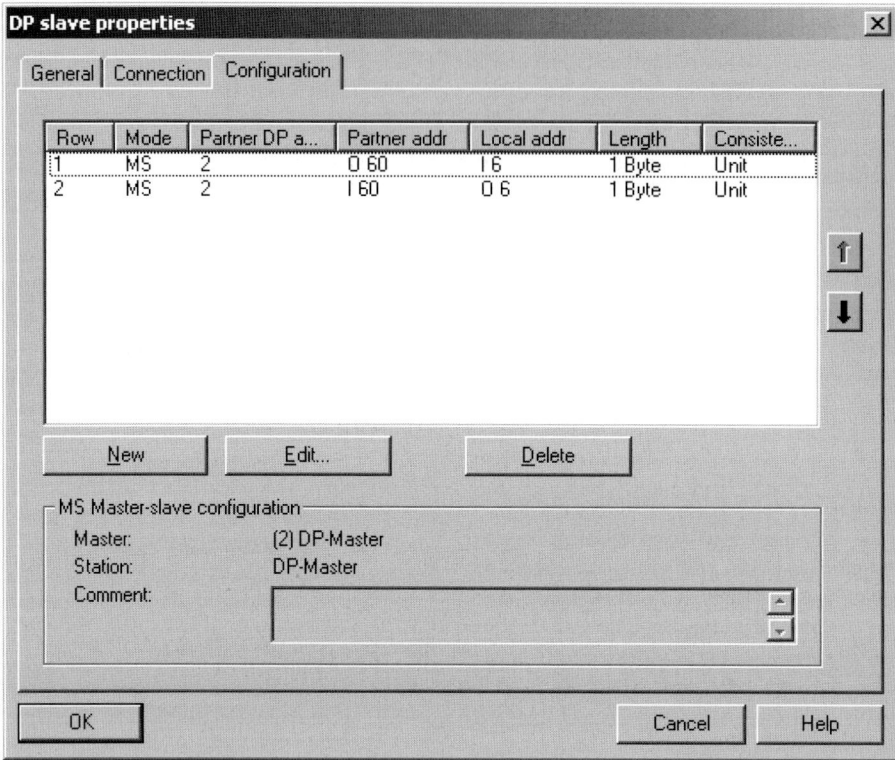

Row	Mode	Partner DP a...	Partner addr	Local addr	Length	Consiste...
1	MS	2	O 60	I 6	1 Byte	Unit
2	MS	2	I 60	O 6	1 Byte	Unit

MS Master-slave configuration
Master: (2) DP-Master
Station: DP-Master
Comment:

Figure 6.35   I/O configuration of I-slave 6

Next, configure a cross communication connection from I-slave 5 to I-slave 6, and vice versa. In the hardware configuration of the DP master, double-click the I-slave 5 to open its "Configuration". Click on "New…" to open the Configuration dialog. In the "Mode" column, select "DX" for cross communication and enter the parameters shown in Figure 6.36. Quit the dialog with OK. The configuration shown in Figure 6.37 then appears. Exit the window with OK.

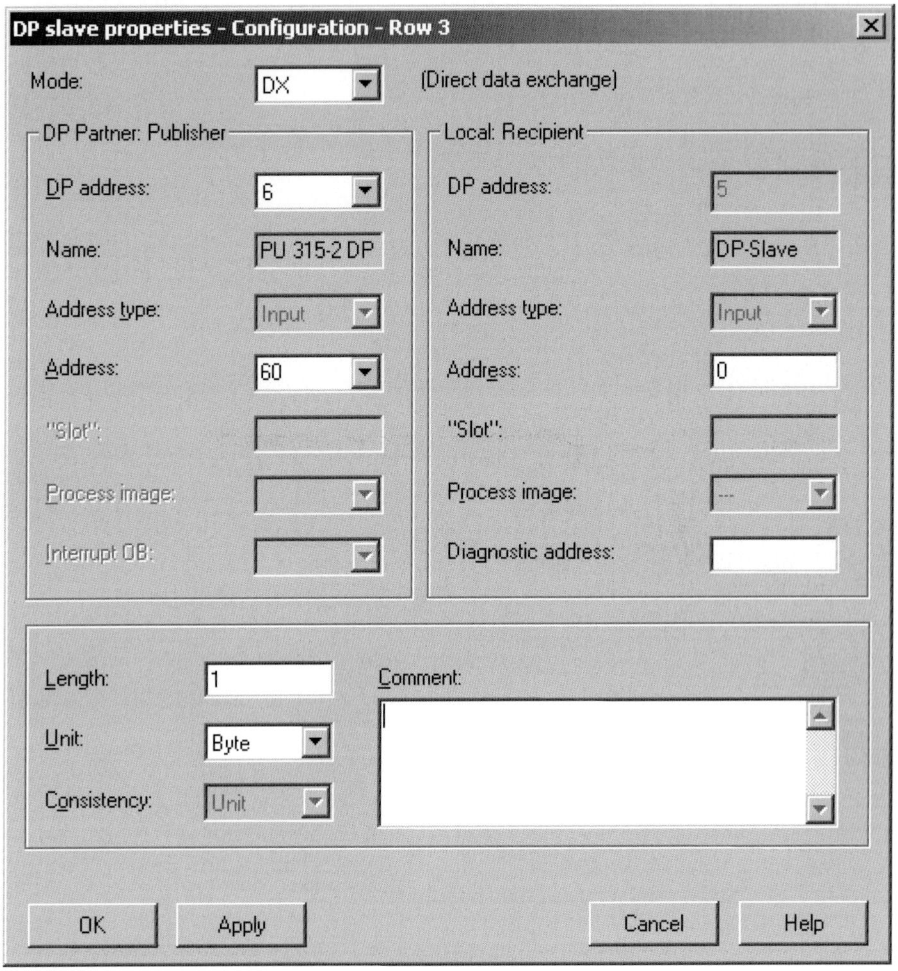

Figure 6.36 Parameter for cross communication from I-slave 5 to I-slave 6

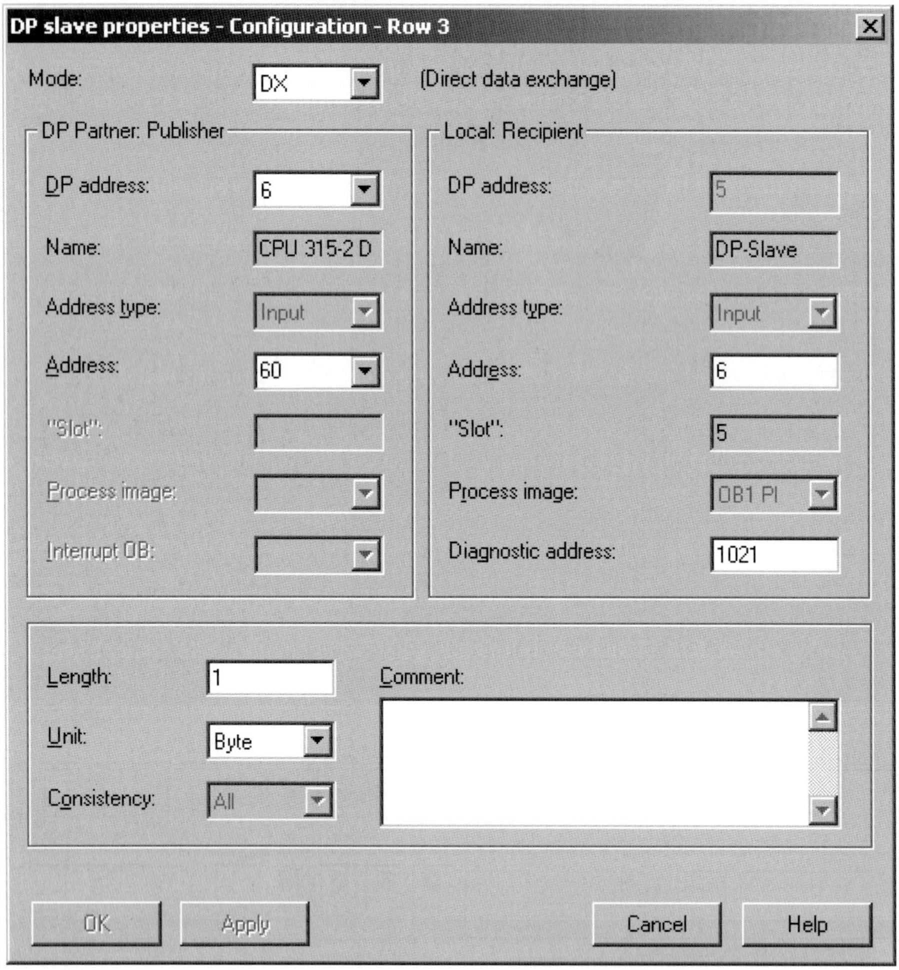

Figure 6.37 Configuring cross communication connection I slave 5 to I slave 6

Use the same procedure for the cross communication connection from I-slave 6 to I-slave 5. Double-click I-slave 6 to open its "DP Slave Properties" dialog box and the "Configuration" tab. Enter the parameters as shown in figure 6.38.

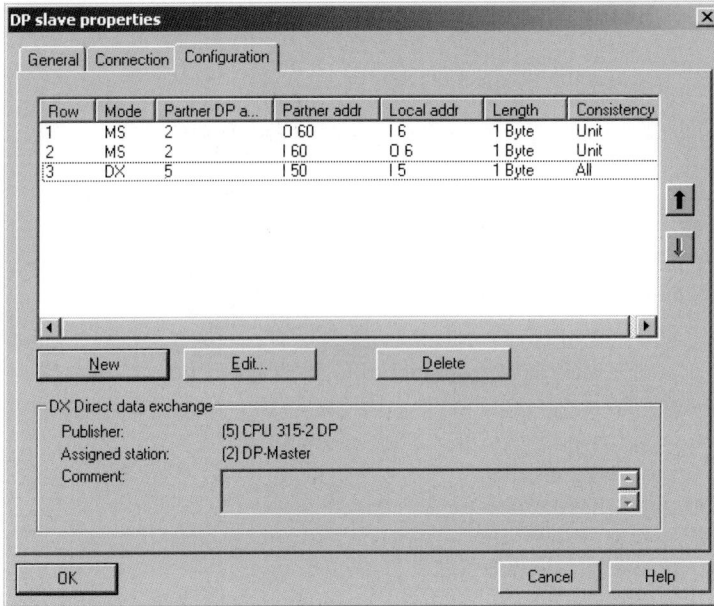

Figure 6.38 Cross communication from I-slave 6 to I-slave 5

Both cross communication connections configured up to now are slave-to-slave connections. Another version of direct communication is the slave-to-master connection. However, in this case the master is not the Class 1 Master responsible for loading the parameter set to the DP slave. Instead, it is simply an additional DP master which is able to receive and further process the current input states of the DP slave.

In our example, we will use the S7-300 station called "DP-Master/Inputs" to implement the slave-to-master connection. Starting from *SIMATIC Manager*, set up the DP master station using the *HW Config* program. Again, use a CPU from the S7-300 series, the CPU 315-2DP. Enter bus address 3 for this DP master, and connect it to the already existing PROFIBUS subnet. Double-click the DP master interface of this station to open the "Configuration" tab of the "Properties – DP Master" dialog box. Enter the two passive cross communication connections for I-slave 5 and I-slave 6 as shown in figure 6.39 ("DX" mode is shown in gray). Quit the dialog box with OK. This DP master has now gained access to the input states of the associated DP slaves. Save and compile the configuration for this station.

Now, load the projects to the individual S7-300 stations. Use load and transfer instructions to implement data communication from and to the configured I/O addresses. Test the data exchange using the STEP 7 function *Monitor/Modify Variables* (see section 6.2.3).

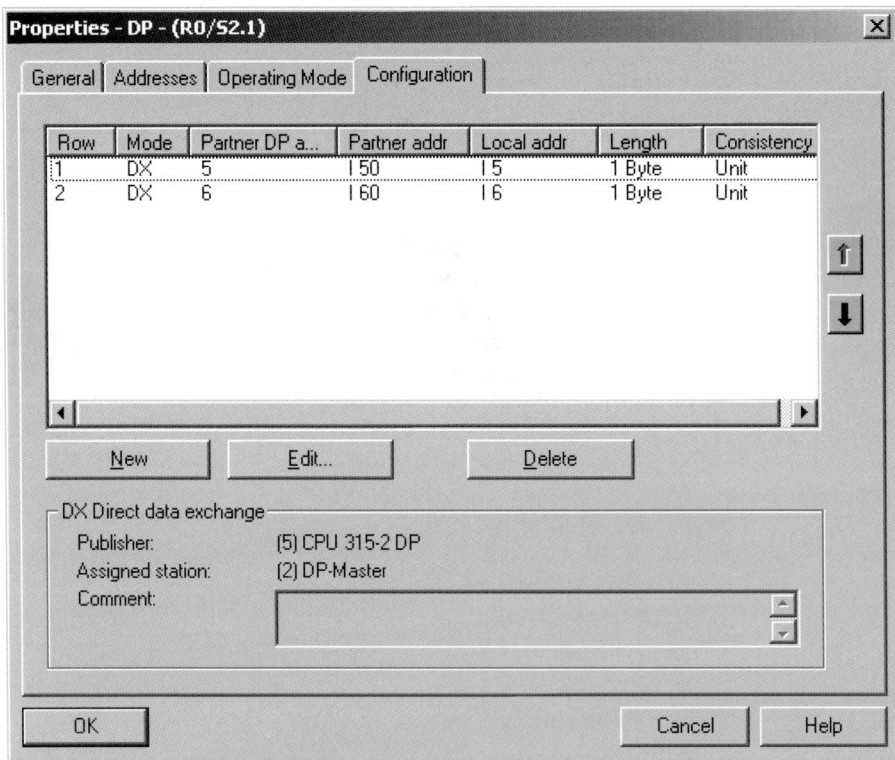

Figure 6.39 Cross communication with "DP master/Inputs"

7 Diagnostic Functions for PROFIBUS DP

Introduction

SIMATIC S7 programmable controllers provide extensive diagnostic facilities for the detection and localization of errors in an automated plant that uses a PROFIBUS DP network. These diagnostic functions can also be employed as monitoring functions. In this case they are automatically executed as part of the user program.

The diagnostic utilities available for a DP network implemented with SIMATIC S7 can be divided into four groups.

▷ Diagnosis by means of LEDs

  Diagnosis using the LEDs on the CPU, the DP master and the individual DP slaves

▷ Diagnosis by means of STEP 7 online functions

  STEP 7 provides a number of online diagnostic functions, such as *Accessible Nodes*, *Diagnose Hardware* and *Module Information*.

▷ Diagnosis by means of the user program

  The S7 DP slaves are totally integrated in the SIMATIC S7 diagnostic concept which provides the user program with appropriate interfaces for fault and failure messages. In addition, a system function can be called in the user program (SFC = system function call) to obtain detailed information on system states and the cause of faults and failures.

▷ Diagnosis using the PROFIBUS monitor

  The PROFIBUS monitor is available for the examination of complex faults or problems with data transmission. This tool is used to record and evaluate telegram communication on PROFIBUS.

This chapter describes the primary diagnostic functions available for the SIMATIC S7 system. It also provides a number of examples on how to include diagnostic interfaces and SFCs in the user program for the evaluation of diagnostic information.

7.1 Diagnosis by means of LEDs on the SIMATIC S7 CPUs, the DP Master Interfaces and the DP Slaves

The front plates of the CPUs of the SIMATIC S7-300 and S7-400 series are equipped with LEDs which indicate the current status of the CPU or the PROFIBUS DP interface. In the event of a system fault, these LEDs give you an initial idea of where to look for the fault.

The LEDs are divided into two groups.

▷ General status and error LEDs for the CPU and

▷ LEDs indicating DP interface faults.

7.1.1 LEDs of the S7-300

General LEDs of the CPU315-2-DP

Table 7.1 lists the general status and error LEDs of the CPU315-2DP of the S7-300 programmable controller. The LEDs are shown in the order in which they are arranged on the CPU's front plate.

Table 7.1 General LEDs of the CPU315-2DP

LED	Meaning	Explanation
SF (red)	Group error	LED lights up for • Hardware errors • Firmware errors • Programming errors • Parameter errors • Calculation errors • Time errors • Faulty memory card • Battery failure or no battery backup at POWER-ON • I/O error (only for external I/O) Remarks: For more accurate fault finding, use a PG programming unit and read the diagnostic buffer of the CPU.
BATF (red)	Battery error	LED lights up if the battery is defective, is missing or is dead.
DC5V (green)	5V DC supply	LED goes on when the internal 5 V DC supply for the CPU and S7-300 bus is okay.
FRCE (yellow)	Reserved	"Force" function is not implemented on this CPU.
RUN (green)	Operational mode RUN	• LED flashes at 2 Hz for at least 3 seconds while the CPU is starting up (CPU startup may be shorter). During CPU startup, the STOP indicator also goes on. The outputs are enabled when the STOP LED goes off. • LED goes on when CPU is in the RUN mode.

Continued on page 181

Table 7.1 Continued

LED	Meaning	Explanation
STOP (yellow)	Operational state STOP	• LED is on when CPU is not processing a user program. • LED flashes at 1-sec intervals when CPU requests overall reset.

LEDs of the DP interface of the CPU 31x-2DP

The meaning of the LEDs for the PROFIBUS DP interface depends on the operating mode of the DP interface. The following two modes exist.

▷ DP master and

▷ DP slave.

LEDs of the CPU 31x-2DP in "DP master" mode

Table 7.2 lists the LED indications of the PROFIBUS DP interface when the CPU 31x-2DP is used as a DP master.

Table 7.2 LEDs of the CPU 31x-2DP in "DP master" mode

SF DP	BUSF	Meaning	Measures
Off	Off	• Configuration okay • All configured slaves can be addressed.	
On	On	• Bus error (hardware fault) • DP interface error • Different baud rates in multi-master operation	• Check the bus cable for short circuit or breaks. • Evaluate the diagnosis information. Define a new configuration, or correct the old one.
On	Flashing	• Station failure • At least one of the assigned slaves cannot be addressed.	Check the bus cable connected to the CPU 31x-2DP. Wait until the CPU 31x-2DP has started up. If the LED does not stop flashing, check the DP slaves, or evaluate the diagnostic information of the DP slaves.
On	Off	Missing or incorrect configuration (this also occurs, when the CPU was not set up as the DP master)	Evaluate the diagnostic information. Define a new configuration, or correct the old one.

LEDs of the CPU 31x-2DP in "DP slave" mode

Table 7.3 lists LED indications of the PROFIBUS DP interface when the CPU 31x-2DP is used as a DP slave.

Table 7.3 LEDs of the CPU315-2DP in "DP slave" mode

SF	DP BUSF	Meaning	Measures
Off	Off	Configuration okay	–
Not relevant	Flashing	The parameter set of the CPU 31x-2DP is incorrect. There is no data communication between DP master and CPU 31x-2DP. Possible causes: • Watchdog monitoring timed out. • Bus communication through PROFIBUS DP is interrupted. • PROFIBUS address defined incorrectly.	• Check the CPU 31x-2DP. • Make sure that the bus plug connector is inserted correctly. • Make sure that the bus cable to the DP master is not interrupted. • Verify the configuration and the parameter set.
Not relevant	On	Bus short circuit	Check the setup of the bus.
On	Not relevant	• Missing or incorrect configuration • No data communication with the DP master	• Check the configuration. • Evaluate the diagnostic interrupt or the diagnostic buffer entry.

7.1.2 LEDs of the S7-400 CPUs Equipped with a DP Interface

Table 7.4 lists the LED indications of the S7-400 CPUs which are equipped with a PROFIBUS-DP interface. The LEDs are shown in the same order in which they are arranged on the CPU.

Table 7.4 LEDs of S7-400 CPUs with DP interface

CPU		DP Interface	
LED	Meaning	LED	Meaning
INTF (red)	Internal error	DP INTF (red)	Internal error on the DP interface
EXTF (red)	External error	DP EXTF (red)	External error on the DP interface
FRCE (yellow)	Forcing	BUSF	Bus error on the DP interface
CRST (yellow)	Complete reset (cold)		
RUN (green)	Oper. state RUN		
STOP (yellow)	Oper. state STOP		

General LEDs of S7-400 CPUs with DP master interface

Table 7.5 explains the LEDs for the status messages of S7-400 CPUs equipped with a DP master interface.

Table 7.5 LEDs for status messages of S7-400 CPUs with DP interface

LED			Meaning
RUN	STOP	CRST	
On	Off	Off	CPU is in operational state RUN.
Off	On	Off	CPU is in operation state STOP. The user program is not processed. Warm or hot restart is possible. If the STOP state was caused by errors, the fault LED (INTF or EXTF) is also switched on.
Off	On	On	CPU is in STOP state. Only a warm restart is possible as next startup mode.
Flashing (0.5 Hz)	On	Off	HOLD state was triggered by test function of the PG.
Flashing (2 Hz)	On	On	A warm start is performed.
Flashing (2 Hz)	On	Off	A hot restart is performed.
Not relevant	Flashing (0.5 Hz)	Not relevant	CPU requests overall reset (cold).
Not relevant	Flashing (2 Hz)	Not relevant	Overall (cold) reset running.

Pending errors or special functions in progress are indicated by the LEDs listed in table 7.6.

Table 7.6 LEDs for errors and special functions of S7-400 CPUs with DP interface

LED			Meaning
INTF	EXTF	FRCE	
On	Not relevant	Not relevant	An internal error (programming or parameter error) was detected.
Off	On	Not relevant	An external error (error not caused by the CPU module) was detected.
Not relevant	Not relevant	On	A PG is performing the "force" function on this CPU. This means that variables of the user program are set to a fixed value and can no longer be changed by the user program.

Table 7.7   LEDs of the S7-400 DP interface

LED			Meaning
DP INTF	DP EXTF	BUSF	
On	Not relevant	Not relevant	An internal error (programming or parameter error) was detected on the DP interface.
Not relevant	On	Not relevant	An external error (error caused by a DP slave and not the CPU module) was detected.
Not relevant	Not relevant	Flashing	One or more DP slaves on PROFIBUS do not respond.
Not relevant	Not relevant	On	A bus error (e.g., cable break or differing bus parameters) on the DP interface was detected.

LEDs of the DP interface of S7-400 CPUs

Table 7.7 lists LED indicators of the PROFIBUS DP interface of S7-400 CPUs.

7.1.3 LEDs of the DP Slaves

DP slave modules are also equipped with LEDs which indicate the operational status and any malfunctions on the DP slave. The number of LEDs and their meaning depend on the type of slave used. For more detailed information, refer to the technical documentation of the particular DP slave.

The LEDs of the DP slaves which we used in our example configuration (section 4.2.5) are described below.

LEDs of the ET 200B 16DI/16DO module

Table 7.8 lists LED indicators of the ET 200B 16DI/16DO module.

Table 7.8   Status and error indicators of the ET 200B 16DI/16DO module

LED	Optical Signal	Meaning
RUN	On (green)	ET200B is in operation (power supply on, STOP/RUN switch in RUN position)
BF	On (red)	• Watchdog monitoring timed out without the station having been addressed (e.g., connection to the S7 DP master has failed). • The station has not yet received its parameter set during commissioning/startup.
DIA	On (red)	For digital 24 V DC output modules, for at least one output: Short circuit or load voltage missing
L1+	On (green)	Voltage for channel group "0" is present (blown fuse or low voltage, typical: +15.5 V, the signal diode goes off).
L2+	On (green)	Voltage for channel group "1" is present (blown fuse or low voltage, typical: +15.5 V, the signal diode goes off).

LEDs of the ET 200M/IM153-2 module

Table 7.9 lists the LED indicators of the ET 200M/IM153-2 module.

Table 7.9   Status and error LEDs of the ET 200M/153-2 module

LED			Meaning	Measures
ON (Green)	SF (Red)	BF (Red)		
Off	Off	Off	No voltage is present, or the IM153-2 has defective hardware.	Check the 24 V DC of the power supply module.
On	Not relevant	Flashing	IM153-2 loaded with incorrect parameter set, or there is no data communication between the DP master and the IM153-2 module. Possible causes: ▷ Watchdog monitoring timed out. ▷ Bus communication through PROFIBUS DP to the IM153-2 module is interrupted.	Check the DP address. Check the IM153-2 module. Make sure that the bus plug connector is inserted correctly. Make sure that the bus cable to the DP master is not interrupted. Turn off and on the on/off switch for 24 V DC on the power supply module. Check the configuration and parameter set.
On	Not relevant	On	Baud rate search or illegal DP address	Set a valid DP address ("1" to "125") on the IM153-2, or check the bus setup.
On	On	Not relevant	Configured setup of the ET 200M module does not match the module's actual setup. Error in an installed S7-300 module, or IM153-2 is defective.	Check the setup of the ET 200M module (module missing or defective, non-configured module installed). Check the configuration. Replace the S7-300 module or the IM153-2.
On	Off	Off	Data communication between DP master and ET 200M is taking place. Defined and actual configuration of the ET 200M match.	

7.2 Diagnosis by means of Online Functions of the STEP 7 Program

The standard STEP 7 programming tool provides a number of online functions for diagnosis. This chapter describes the diagnostic functions and gives an example of how to use these functions in a PROFIBUS DP system.

7.2.1 *Display Accessible Nodes* in *SIMATIC Manager*

In *SIMATIC Manager*, you can call the *PLC → Display Accessible Nodes* function to check which active and passive bus nodes are connected to the MPI or PROFIBUS network. You can also use this function if you want to carry out an error diagnosis for the MPI and PROFIBUS stations connected to the network, but you have no STEP 7 data for these stations.

Before you can use this online diagnostic function, adjust the PG/PC interface to the baud rate set on the PROFIBUS network (default of 187.5 kBaud for MPI) and the selected bus profile. When the function is started, the online interface of the PG/PC is passive on the bus and checks to determine whether the baud rate set on the interface matches the one set on the PROFIBUS network. If this is not the case, an appropriate error message is displayed. The same applies when a bus station address is assigned twice by the connected PG/PC. The PG/PC does not report as an active bus station and is not included in the token ring until it has verified that the baud rates match and no bus station address is assigned twice.

A maximum transmission speed of 1.5 MBaud can be set with an MPI/ISA card. Diagnosis at higher baud rates requires an additional interface card such as a CP 5411 (ISA card), CP 5511 (PCMCIA card) or CP 5611 (PCI card). All these interfaces are fully supported by the standard STEP 7 package. This means that no additional drivers are required.

Start the diagnostic function in *SIMATIC Manager*. In the menu bar, select *PLC → Display Accessible Nodes*. A dialog box appears showing all programmable modules, such as CPUs, FMs and CPs, that can be addressed on the network. Their MPI or bus addresses are also displayed. The dialog box also lists MPI and bus stations which were not configured with STEP 7 (e.g., Operator Panels). The bus station connected directly to the PG programming unit or PC through the MPI cable or the active bus cable is indicated by the note "direct" after the bus address (see figure 7.1).

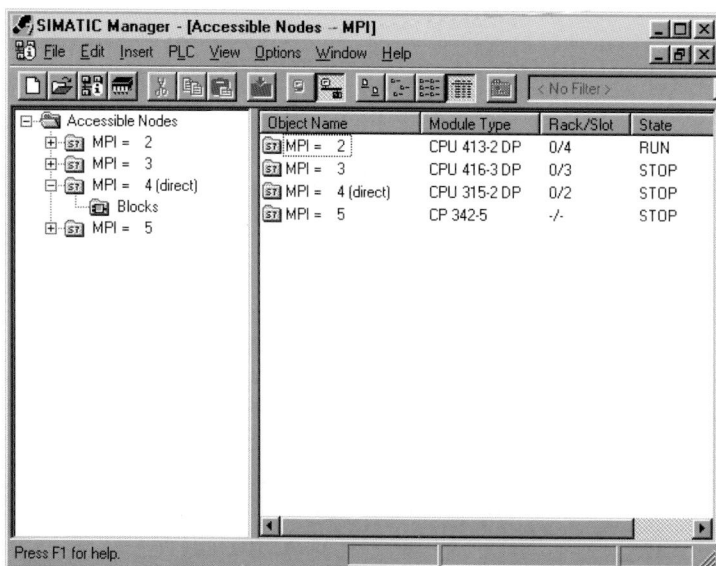

Figure 7.1 *Display Accessible Nodes* function through MPI

This diagnostic function provides quick access to programmable modules, which is particularly important during service and maintenance.

However, remember that changes in the Online view (e.g., station failure) are not automatically updated in the *Accessible Nodes* dialog box opened on the screen. To update the contents of this dialog, press function key "F5" or, in the menu bar, select VIEW → UPDATE.

A right click on an MPI station opens the shortcut menu. Select PLC to open another submenu. The following menu commands are part of the diagnostic function:

▷ MONITOR/MODIFY VARIABLES. This command starts the STEP 7 function *Monitor/ Modify Variables* which allows you to set and monitor variables of the destination system without having a project for it.

▷ OPERATING MODE. You can use this function to scan the current status of the station and, if possible, change it.

▷ MODULE INFORMATION. See chapter 7.2.3.

▷ DIAGNOSE HARDWARE. See chapter 7.2.4.

Setting the PG/PC online interface

In *SIMATIC Manager*, select OPTIONS → SET PG/PC INTERFACE... (see figure 7.2). Use a PG740 programming unit with integrated MPI card. In the *Interface parameter set used* group, select "MPI-ISA on Board (PROFIBUS)." Press the PROPERTIES... button to view the details of this parameter set, and select an unassigned PROFIBUS address for the PG programming unit.

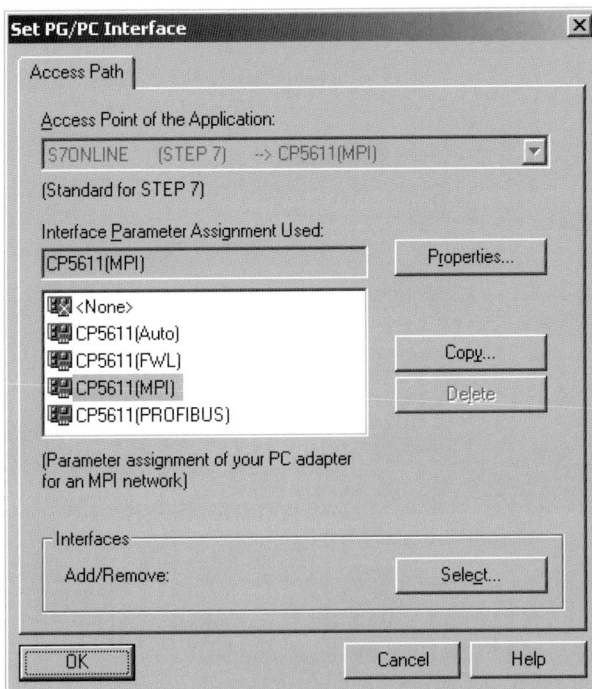

Figure 7.2
Setting the PG/PC interface

Set the baud rate to the value actually used by your system, and compare the highest station address and the profile for the bus parameters to be used with the values set on the system. Acknowledge your settings with OK.

Connect the MPI/DP interface of your PG to PROFIBUS. However, remember that an active cable (PROFIBUS line with integrated repeater) must be used to connect a PG to PROFIBUS. Otherwise, you may cause a bus malfunction when you plug in the cable.

Once the PG/PC device is physically connected to PROFIBUS, click the ACCESSIBLE NODES button in the toolbar of *SIMATIC Manager*, or select PLC → DISPLAY ACCESSIBLE NODES from the menu bar. The PG device now "listens in" the bus and generates a "life list" of all PROFIBUS devices connected to the bus. When the life list is complete, the stations are displayed in *SIMATIC Manager*. The type of station is also shown (i.e., active station (DP master) or passive station (DP slave)). If the PG programming unit is connected to the PG socket of the PROFIBUS plug connector of a station, then this is indicated by the note "direct" after the PROFIBUS address of the station (see figure 7.3).

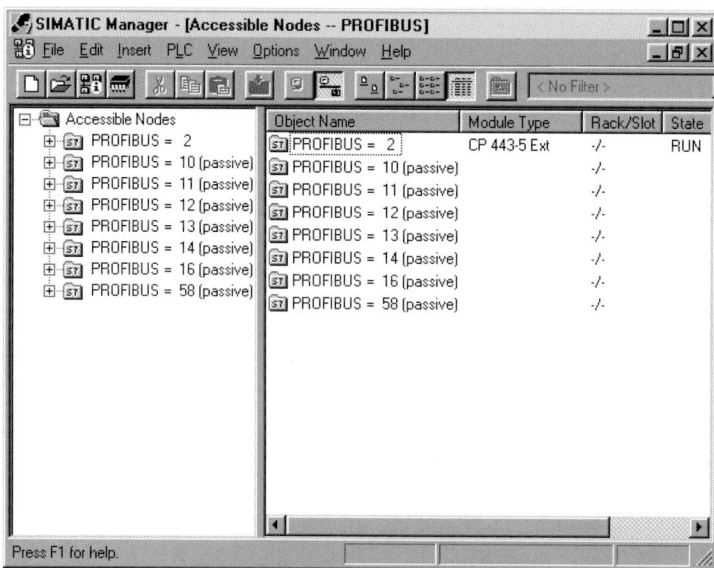

Figure 7.3 *Accessible nodes* function through PROFIBUS

For example, the "accessible nodes" function can be used to check the PROFIBUS addresses set on the DP slaves or when a cable break in the PROFIBUS line is suspected. You can then check whether it is still possible to address any of the modules and, if so, which modules these are.

Further error diagnosis of the connected PROFIBUS stations is only available if the selected station supports STEP 7 diagnostic functions. For example, S7 CPUs with PROFIBUS DP interface support these diagnostic functions.

Click the PROFIBUS address of a CPU to open the shortcut menu. Alternatively, you can select a diagnostic function from the PLC menu. It provides the functions *Monitor/Modify Variables, Module Information..., Operating Mode, Diagnose Hardware, etc.*

In *SIMATIC Manager*, double-click the PROFIBUS address of an accessible CPU to open the object. The *Blocks* folder of the CPU appears. Similarly, double-click the *Blocks* folder to view the user blocks in the right-hand half of the window. You can now open these user blocks, modify them and download them to the CPU. However, you cannot use symbolic programming at this point because this would require the STEP 7 offline project.

7.2.2 *ONLINE* Function in *SIMATIC Manager*

If you have the STEP 7 project of a system configuration, you can use the online diagnostic function of the STEP 7 program to open STEP 7 blocks with symbolic names while the system is running. In *SIMATIC Manager*, select the ONLINE button from the toolbar, or select VIEW → ONLINE in the menu bar. This changes the view of your project from the offline mode to the online mode. Although the online diagnostic function is usually applied in MPI networks, you can also use it when your PG programming unit or PC is connected through PROFIBUS. To do this, open the project, and set your PG/PC interface to the values valid for your system as described in chapter 7.2.1. When you then gain access to the target system you can decide whether or not the hardware configured in the project shall be taken into account.

To access the system with configured hardware, open the project, and in the menu bar select VIEW → ONLINE. This opens the online view of the station. Then double-click the station for which you want to get the online view, to list the programmable modules it contains. A dialog box automatically appears in which you can define the connection properties, such as the PROFIBUS address of the selected station and its slot (see figure 7.4). Enter these properties of the station or the CPU you want to investigate, and close the

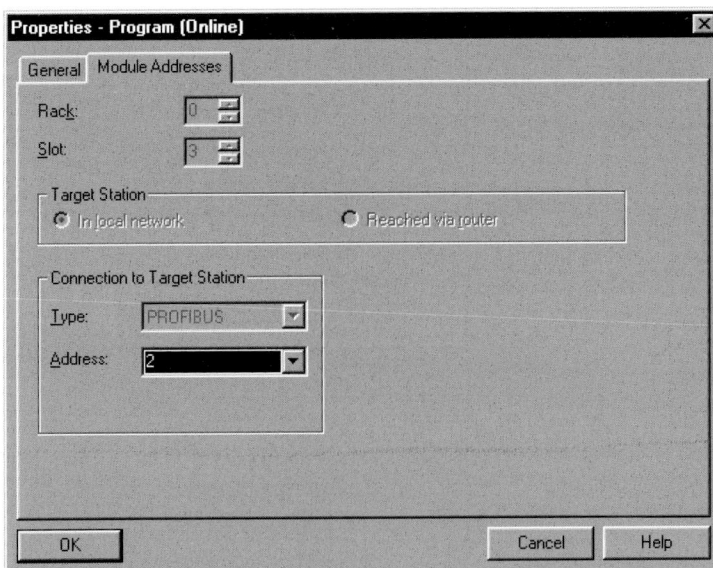

Figure 7.4 Properties of the connection

189

dialog box with OK. This dialog box only appears for the first request of online access. The information you enter here is stored in the STEP 7 project, so you will not be asked for it in subsequent requests for online access. Double-click the module of the opened station which you want to investigate. This establishes the connection to it using the entered settings. You can now carry out an online diagnosis of the entire S7 station or STEP program through the PROFIBUS connection.

You can also request access to the system without considering the configured hardware. This means that you don't use the hardware configuration from the offline project. Again, open the online view by selecting VIEW → ONLINE from the menu bar, or the ONLINE button from the toolbar. Select the S7 program listed directly under the project name, and right-click to open the shortcut menu. Select OBJECT PROPERTIES... to open the PROPERTIES-PROGRAM (ONLINE) dialog box. Change to the MODULE ADDRESSES tab and enter the PROFIBUS address of the CPU which you would like to investigate. Close the dialog box with OK. The connection to this CPU is established, and you can now test the STEP 7 program in the online mode.

7.2.3 *Module Information* in *SIMATIC Manager*

This diagnostic function provides actual module information. The scope of this information depends on the type of module selected. The dialog box that displays the module information is made up of a number of tabs. It only displays those tabs that are relevant to the selected module. In addition to the information offered on a varying number of tabs, the dialog box contains some permanent information, such as the operating mode of the module. If you have not selected an S7 CPU, the dialog box displays the status of the selected module seen from the CPU. This could be for instance *OK, Error, Module does not exist, etc.*

Table 7.10 shows which tab pages the "Module Information" dialog box offers for the individual module types.

For example, FM modules (*Function Module*) have system diagnostic capabilities. Most analog SM modules only support simple diagnostic functions. Most digital SM modules do not support any diagnosis at all.

There are several ways to open the "Module Information" dialog box.

▷ Using the *Accessible Nodes* function in *SIMATIC Manager*. Right-click the target system which you want to investigate. This opens the shortcut menu. Select PLC → MODULE INFORMATION.

Table 7.10 Information on the module types and the relevant tabs in the "Module Information" dialog box

Tab Page	CPU or M7-FM	System Diagnosis	Diagnosis	No Diagnosis	Standard DP Slave
General	X	X	X	X	X
Diagnostic Buffer	X	X			
Memory	X				
Scan Cycle Time	X				

Continued on page 191

Table 7.10 Continued

Tab Page	CPU or M7-FM	System Diagnosis	Diagnosis	No Diagnosis	Standard DP Slave
Time System	X				
Performance Data	X				
Stacks	X				
Communication	X				
Diagnostic Interrupt		X	X		
DP Slave Diagnosis					X

▷ Using the *Online* function in *SIMATIC Manager*. Select VIEW → ONLINE to change the view of your project from offline to online. Display the station you want to investigate in the left-hand half of the *SIMATIC Manager* screen. Double-click the station to open it, and right-click the programmable module or the CPU to open the shortcut menu. Select PLC → MODULE INFORMATION ...

▷ Using the *Diagnose hardware* function in *SIMATIC Manager*. See chapter 7.2.4.

Figure 7.5 shows the "General" tab of the *Module Information* dialog box. The individual tabs provide different information. Table 7.11 lists the available tab pages of this dialog box and their purpose and contents. In practice you will always only see those tabs that are applicable to the particular module which you have selected for diagnosis.

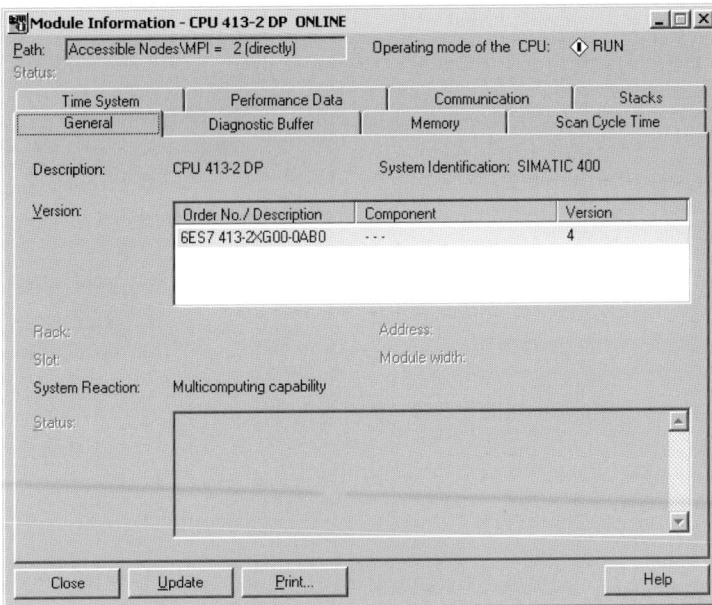

Figure 7.5 "Module Information" dialog box

Table 7.11   Contents and purpose of the tabs of the *Module Information* dialog box

Tab Title	Contents	Use
General	ID information of the selected module (e.g., type, version, order no., rack, slot)	Compare the online ID information of the installed module with the ID information of the module you have configured in *HW Config*.
Diagnostic Buffer	Overview of events in the diagnostic buffer	Evaluate what caused a CPU to STOP
Memory	Current load of work memory and load memory of the selected CPU or M7-FM module	Check the memory utilization before you transmit new or expanded blocks to a CPU
Scan Cycle Time	Duration of the shortest, longest and last cycle of the selected CPU or M7-FM module	Use this information to check the minimum cycle time which you have defined in the configuration, and the maximum and current cycle time.
Time System	Current time of day, hours of operation, and information on clock synchronization	Check the time of day and date of a module. Check the time synchronization
Performance Data "Blocks" (can be called from the "Performance Data" tab)	Memory configuration, address areas and available blocks of the selected CPU/FM module Indication of all types of blocks which are included in the function scope of the selected module. List of the OBs, SFBs and SFCs which can be used with this module	Use this information prior to and during generation of a user program. Check whether an existing user program is compatible with a special module.
Communication	Baud rates, connection overview, communication load and maximum size of the telegrams	Use this information to determine which and how many connections of the CPU or M7-FM are possible or occupied.
Stacks	Contents of B stack, I stack and L stack. From here you can also change to the block editor.	Use this information to determine the cause of a transition to STOP and to correct a block
Diagnostic Interrupt	Diagnostic information of the selected module	Determine the cause of a module fault
DP Slave Diagnostics	Diagnostic information of the selected DP slave in acc. with EN 50 170	Determine the cause of an error of the DP slave

The following information is shown on each tab page.

▷ ONLINE path to the selected module
▷ Operating mode of the related CPU (e.g., RUN, STOP)
▷ Status of the selected module (e.g., Fault, OK)
▷ Operating mode of the selected module (e.g., RUN, STOP) if these have their own operating mode (e.g., IM 467).

Each time you change from one tab to another in the *Module Information* dialog box, new data is read from the module and the contents of the dialog box are updated. However, the contents of a tab are not automatically updated while the tab is open. Press the "Update" button if you want to refresh the displayed information without changing the tab.

We will now describe in more detail the most important tabs of the "Module Information" dialog box.

Diagnostic Buffer

The "Diagnostic Buffer" tab reads the contents of the diagnostic buffer of the module you want to investigate, provided that the module supports system diagnosis, such as a CPU. All diagnostic events and related diagnostic information are registered in the diagnostic buffer in the order in which they occurred (see figure 7.6). The contents of the diagnostic buffer are retained even when the CPU is completely reset.

Diagnostic events are for instance module errors, a system error on the CPU, changes in the operating mode (e.g., from RUN to STOP), and errors in the user program.

The information stored in the diagnostic buffer allows long-term analysis of system errors. You can use this information to determine the cause of a STOP or to trace back the occurrence of individual diagnostic events even long after the actual event occurred.

To obtain additional information on a particular event listed in the diagnostic buffer, select this event and press the "Help on Event" button. Diagnostic entries that refer to an error location (block type, block number and relative address) point to the related block.

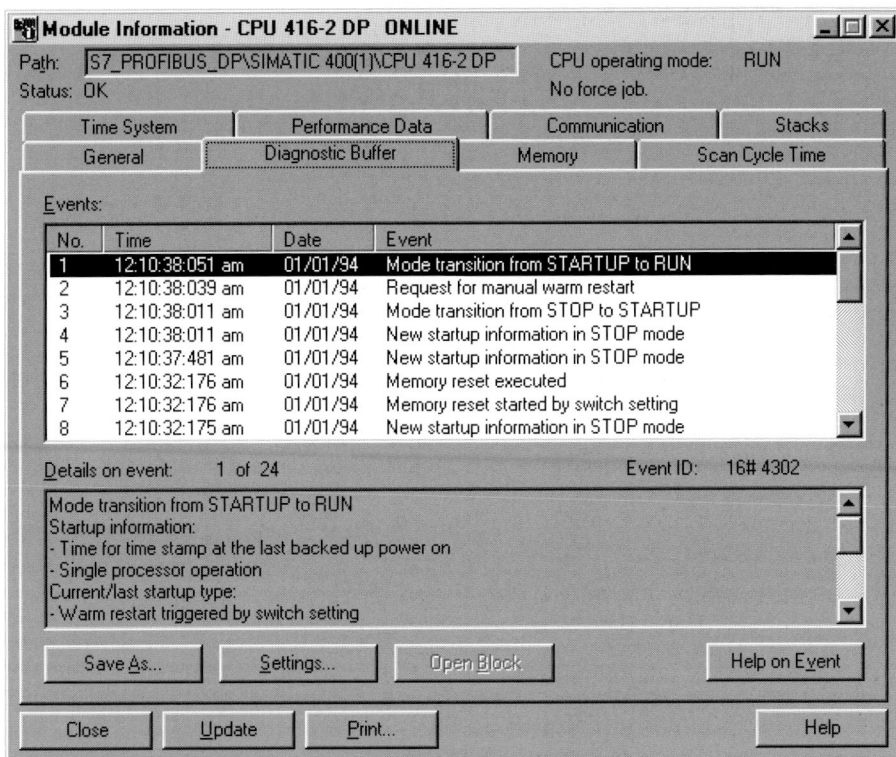

Figure 7.6 "Diagnostic Buffer" tab in the "Module Information" dialog box

193

To open the block, select the entry and press the "Open Block" button. In the opened block, the cursor indicates the point in the program which caused the event.

The diagnostic buffer is a ring buffer. The maximum number of entries may vary with the module selected. When the maximum number of entries is reached and a new diagnostic buffer event arrives, the oldest entry at the end of the list is deleted and all entries are shifted down by one position. Therefore you will always see the most recent diagnostic entry at the top of the list.

Diagnostic Interrupt

The "Diagnostic Interrupt" tab shows information about module faults, provided that the module concerned supports diagnostic functions. The "Standard diagnostics module" group lists internal and external module faults and related diagnostic information (see figure 7.7). Here are some examples of what can be displayed on the "Diagnostic Interrupt" tab.

▷ Module failed.

▷ Channel error.

▷ External auxiliary voltage missing

▷ Module not loaded with parameter set

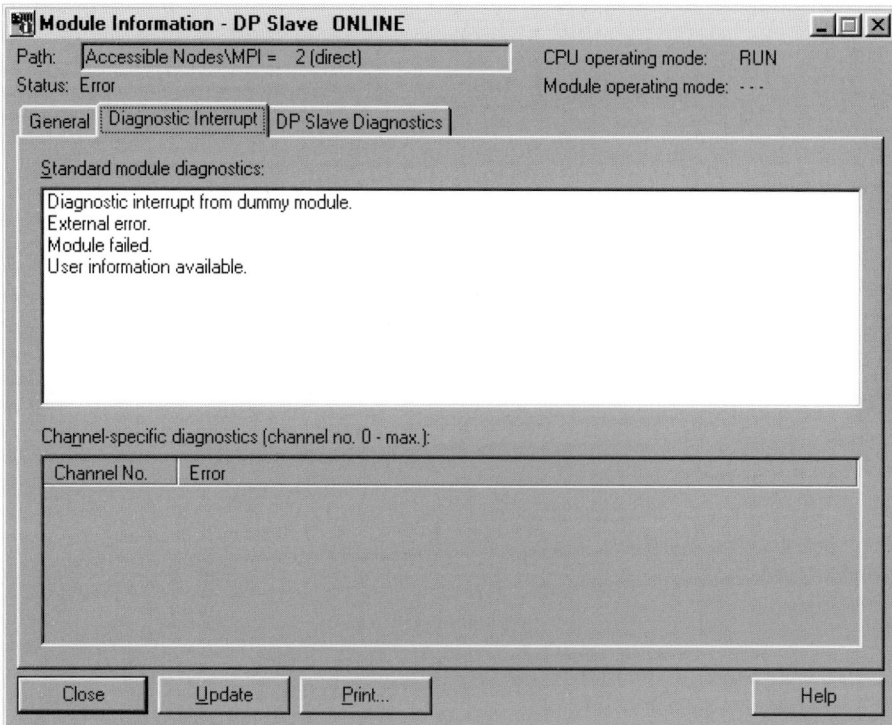

Figure 7.7 "Diagnostic Interrupt" tab

Diagnostic information about channel errors are shown in the "Channel-specific diagnostics" group. Again, here are some examples of this type of information.

▷ Configuration/parameter error

▷ Wire break

▷ Reference channel error

DP Slave Diagnostics

The "DP Slave Diagnostics" tab provides diagnostic information about DP slaves. This information is represented in accordance with the EN 50 170 standard (see figure 7.8).

The "Standard slave diagnostics" group shows general and device-related diagnostic information about the slave.

▷ General diagnostic information about the DP slave

This type of information refers to correct startup or failure of the DP slave. Error messages such as "Slave cannot be addressed", configuration errors or parameter errors are shown in particular here.

▷ Device-related diagnostic texts about the DP slave

These device-related diagnostic texts are derived from the GSD file (device master file). If the diagnostic message is not available in the GSD file, the diagnostic information cannot be provided in the form of a plain-text message.

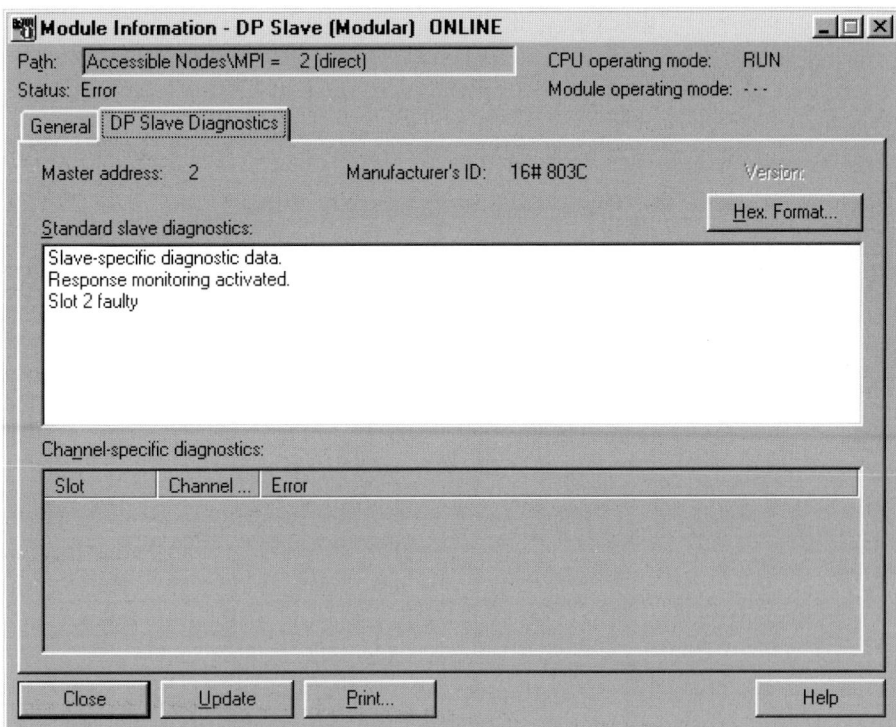

Figure 7.8   "DP Slave Diagnostics" tab

The "Channel-specific diagnostics" group displays channel-related diagnostic texts for configured modules of the DP slave. For each diagnostic message listed in this group, the channel which caused the message is indicated. The channel is unambiguously identified by the slot of the module and the channel number.

Device-related diagnostic texts are derived from the GSD file (device master file). If the GSD file does not contain the diagnostic message, the diagnostic information cannot be provided in the form of a plain-text message. Press the "Hex Format..." button if you want to see the complete diagnostic telegram in hexadecimal format.

7.2.4 *Diagnose Hardware* in *SIMATIC Manager*

There are several ways to call the "Diagnose Hardware" function.

▷ Using the *Accessible Nodes* function in *SIMATIC Manager*. Right-click the target system which you want to investigate. This opens the shortcut menu. Select PLC → DIAGNOSE HARDWARE.

▷ Using the *ONLINE* function in *SIMATIC Manager*. Select VIEW → ONLINE to change the view of your project from offline to online. Right-click the station you want to investigate to open the shortcut menu. Select PLC → DIAGNOSE HARDWARE.

The "Diagnosing Hardware – Quick View" dialog box appears. Icons in the "Module" column indicate the status of the module. For example, if a DP slave is faulty, the quick view displays an icon for the DP slave in addition to displaying the CPU icon (see figure 7.9). Table 7.12 describes the icons. Module faults are only detected and indicated by the module status icon if the module concerned supports diagnostic functions or the diagnostic interrupt has been enabled.

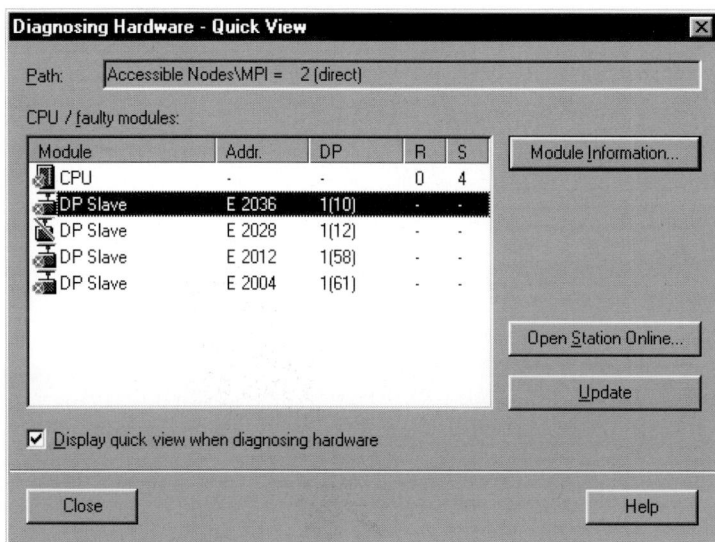

Figure 7.9 "Diagnosing Hardware – Quick View" dialog box

Table 7.12   General description of the diagnostic icons

Diagnostic Icon	Meaning
Red diagonal bar in front of the module icon	Configured and actual configurations don't match. The configured module does not exist, or a different module type is installed.
Red dot with white cross	Module is faulty. Possible causes: Detection of a diagnostic interrupt or an I/O access error.
Low-contrast representation of the module	No diagnosis is possible because no online connection exists or the CPU does not supply diagnostic information on the module (e.g., power supply, sub-modules).
Red vice around the module	"Force variables" is being performed on this module (i.e., variables in the module's user program are preset to fixed values which cannot be changed by the program). The symbol for forcing can also be found in connection with other icons.

The "Diagnosing Hardware – Quick View" dialog box offers a number of buttons for additional functions (see figure 7.9). The button "Module Information..." takes you directly to the "Module Information" tab described earlier in this chapter. The "Update" button refreshes the contents of the "Diagnosing Hardware – Quick View" dialog box. The "Open Station Online ..." button loads the hardware configuration of the selected station. During this load procedure, every module that has been configured is checked. Incorrect or faulty modules are indicated by the relevant icons (see figure 7.10). To obtain additional diagnostic information about a module, right-click the module to open the shortcut menu, and select MODULE INFORMATION ....

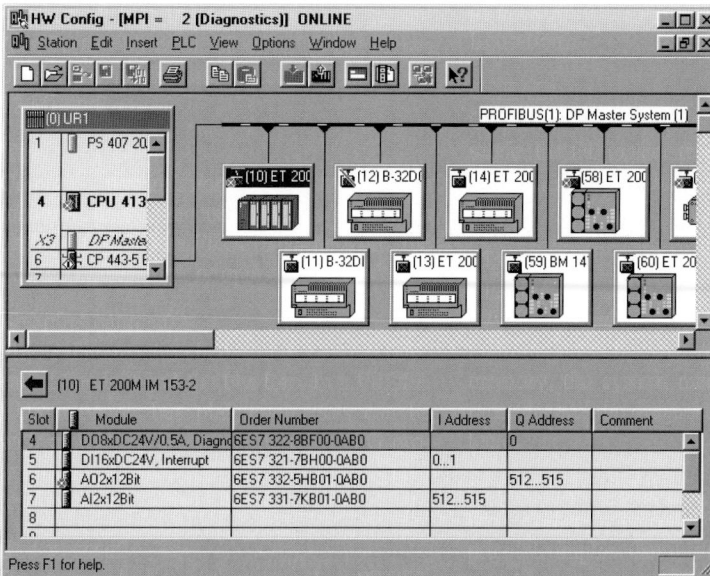

Figure 7.10   Configuration loaded by means of the "Diagnose Hardware" function

7.3 Diagnosis by means of the User Program

SIMATIC S7 programmable controllers offer a variety of diagnostic functions which can be executed by the user program. Systematically applied, these diagnostic functions determine the exact cause of a system fault so that your user program can react accordingly.

In the following we will discuss only some of the diagnostic functions available. The examples given relate to the example project developed earlier in this book (see section 4.2.5).

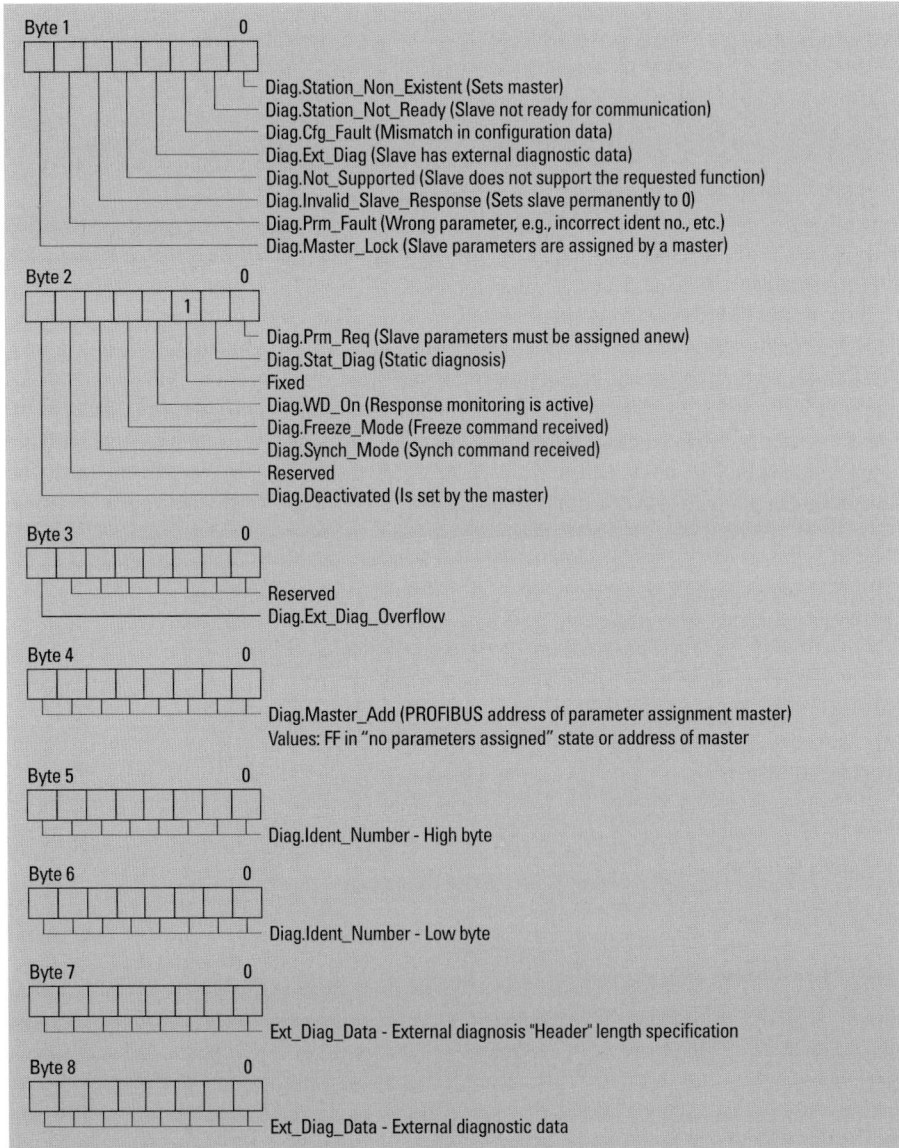

Byte 1

- Diag.Station_Non_Existent (Sets master)
- Diag.Station_Not_Ready (Slave not ready for communication)
- Diag.Cfg_Fault (Mismatch in configuration data)
- Diag.Ext_Diag (Slave has external diagnostic data)
- Diag.Not_Supported (Slave does not support the requested function)
- Diag.Invalid_Slave_Response (Sets slave permanently to 0)
- Diag.Prm_Fault (Wrong parameter, e.g., incorrect ident no., etc.)
- Diag.Master_Lock (Slave parameters are assigned by a master)

Byte 2

- Diag.Prm_Req (Slave parameters must be assigned anew)
- Diag.Stat_Diag (Static diagnosis)
- Fixed
- Diag.WD_On (Response monitoring is active)
- Diag.Freeze_Mode (Freeze command received)
- Diag.Synch_Mode (Synch command received)
- Reserved
- Diag.Deactivated (Is set by the master)

Byte 3

- Reserved
- Diag.Ext_Diag_Overflow

Byte 4

- Diag.Master_Add (PROFIBUS address of parameter assignment master)
  Values: FF in "no parameters assigned" state or address of master

Byte 5

- Diag.Ident_Number - High byte

Byte 6

- Diag.Ident_Number - Low byte

Byte 7

- Ext_Diag_Data - External diagnosis "Header" length specification

Byte 8

- Ext_Diag_Data - External diagnostic data

Figure 7.11 General representation of the diagnostic data in accordance with the EN 50 170 standard

7.3.1 DP-Slave Diagnosis Using SFC13 *DPNRM_DG*

The system function *DPNRM_DG* called by SFC13 reads standard diagnostic data of a DP slave. Contents and representation of the supplied information comply with the EN 50 170 standard.

The maximum telegram length which can be read with SFC13 is 240 bytes although the EN 50 170 standard permits a maximum telegram length of 244 bytes. If the diagnostic data telegram is too long, the overflow bit is set. This overflow bit is part of the data telegram read from the slave. Figure 7.11 illustrates the general layout of the diagnostic data.

System function SFC13 can be called in the cyclic program (OB1), in the diagnostic interrupt OB (OB82) and in the OB for station failure and recovery (OB86). Remember that SFC13 reads the diagnostic data asynchronously. This means that the read procedure requires repeated calls of the system function after being triggered (REQ = "1") to completely read the diagnostic data of a DP slave and enter it in the target area specified in the RECORD parameter.

In a fault or failure situation handled by organization blocks OB82 or OB86, it is important that the diagnostic data read from the DP slave reflects the most recent status. We therefore recommend that you call SFC13 repeatedly in a loop until the output parameters of the system function indicate successful completion of the read procedure.

Figure 7.12 shows how system function SFC13 is called in OB82 to find out the cause of a faulty ET200B 16DI/16DO module. The program evaluates the coming interrupt and the going interrupt separately and writes this information to two different data areas. SFC13 continues in a loop until the BUSY parameter indicates that the job has been executed. Figure 7.13 shows the principle of operation of system function SFC13.

To test the example program, set up DB13 with a minimum length of 132 bytes and call SFC13 in OB82 as shown in figure 7.13. To do this, start *SIMATIC Manager*, and open the example project S7_PROFIBUS_DP developed earlier (see section 4.2.5). Check the hardware configuration of the S7-400 CPU again. Only the ET200B 16DI/16DO module should be connected to the DP master branch. Clear the CPU completely by means of an overall reset, set the operating mode switch of the CPU416-2DP to STOP, and transfer the configuration to the CPU. Using a PROFIBUS cable, connect the DP interface of the CPU to the PROFIBUS interface of the ET200B module. Turn the key-operated switch of the CPU to RUN-P. The CPU switches to RUN mode, and all error LEDs go off. In *SIMATIC Manager*, open the *Blocks* folder of the CPU416-2DP, and in the shortcut menu opened with a right-click, select INSERT NEW OBJECT → ORGANIZATION BLOCK. In the subsequent dialog box, enter "OB82" and confirm with OK. This inserts an empty block shell for OB82 in the *Blocks* folder. Double-click OB82 to open it. This automatically starts the STEP 7 tool *LAD/STL/FBD....... S7 Program*. Enter the example program as shown in figure 7.12. Save OB82, and transfer it to the CPU using the *Download* button from the toolbar, or the command PLC → DOWNLOAD from the menu bar. Set the operating mode switch of the CPU416-2DP to RUN-P, and change to STATUS using the relevant button from the toolbar.

```
    L        #OB82_EV_CLASS                    //Load event
    L        B#16#39                           //Check for "coming" interrupt
    ==I
    JC       go1
//
//Read program section for "going" diagnostic interrupt
//
go2: CALL SFC13
         REQ     : =TRUE
         LADDR   : =W#16#1FFC                  //Diagnostic address of the ET200B station
         RET_VAL : =MW240
         RECORD  : =P#DB13. DBX 100. 0 BYTE 32
         //Diagnostic data DB13 starting at DBB100
         BUSY    : =M230. 0

    A M  230. 0
    JC go2
    BEA
//
//Read program section for "coming" diagnostic interrupt
//
go1: CALL SFC13
         REQ     : =TRUE
         LADDR   : =W#16#1FFC                  //Diagnostic address of the ET200B station
         RET_VAL : =MW240
         RECORD  : =P#DB13. DBX 0. 0 BYTE 32
         //Diagnostic data DB13 starting at DBB0
         BUSY    : =M230. 0

    A       M         230. 0
    JC      go1
```

Figure 7.12 Calling SFC13 *DPNRM_DG* in OB82

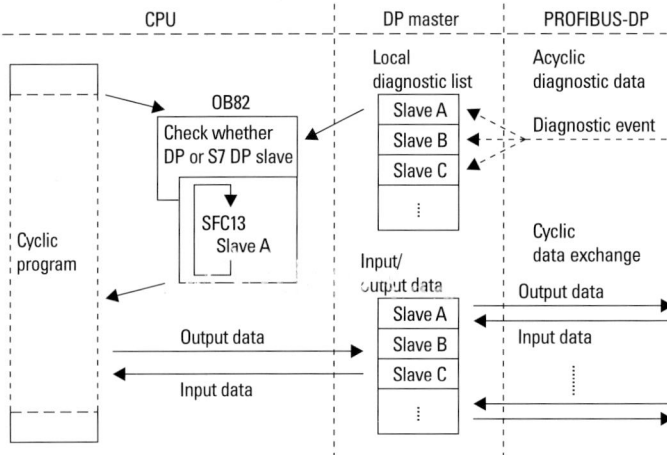

Figure 7.13 Principle of operation of the example program with SFC13 *DPNRM_DG* in OB82

To create a fault on the ET 200B 16DI/16DO module, disconnect the load voltage of a channel group from the 24 V DC power supply. OB82 is called, and system function SFC13 starts evaluating the diagnostic information. Progress and status of this procedure are indicated in the STL editor. You may now start the *Monitor/Modify Variables* function to analyze the diagnostic data.

7.3.2 Diagnosis Using SFC51 *RDSYSST* in OB82

S7 DP slaves or S7-300 modules on an S7 DP slave offer extended diagnostic functions. An S7 DP slave of modular design allows precise error diagnosis on each individual S7-300 module. The ET 200M unit containing several S7-300 modules is an example of such a modular S7 DP slave. These PROFIBUS components are able to send a diagnostic interrupt to the DP master or the CPU, which calls organization block OB82. Inside OB82, system function SFC51 *RDSYSST* is called to carry out an extended error diagnosis.

SFC51 is an asynchronous system function. This means that it must be called more than once to completely read the diagnostic information and write it to the target data area specified in the DR parameter. Synchronous execution of SFC51 is also possible. For this, SFC51 must be called in OB82 to read data record "0" or "1" of the module which is the cause of the interrupt. This type of error diagnosis filters the information fetched from the DP slave so that only those details are read that actually relate to the fault.

Use system function SFC51 if you want to focus the error diagnosis on the affected S7 DP slave or S7-300 module. SFC51 reads data record "0" (4 bytes) or data record "1" (16 bytes). Contents and layout of the read data record are the same as for the diagnostic data of a module that has been connected locally in the central or expansion rack. SFC51 can thus be used if centrally and decentrally connected modules must be diagnosed in the same manner.

The local data supplied by OB82 allows you to program SFC51 with variable calls. In this way you don't need to program a separate SFC51 call for every S7 DP slave, or for every S7-300 module used on an S7 DP slave.

The example program shown in figure 7.14 reads data record "1" of the faulty module which generated the diagnostic interrupt. The program distinguishes between a "coming" event and a "going" event. The diagnostic information fetched by the program can then be evaluated more accurately in OB82 or in the cyclic program (OB1).

In our example, the local data of OB82 is applied to SFC51. The local variable OB82_-EV_CLASS (event class and identifiers) has the following meaning.

- ▷ Going event   B#16#38
- ▷ Coming event   B#16#39

The local variable OB82_IO_FLAG (type of module) supplies the following values.

- ▷ Input module   B#16#54
- ▷ Output module   B#16#55

The SFC51 call in OB82 requires the variable structure "SZL_HEADER" as shown in table 7.13. Therefore, add variable "SZL_HEADER" to the local data of OB82.

Table 7.13   Variable structure "SZL_HEADER"

Name	Type
SZL_HEADER	STRUCT
LENGTH_DR	WORD
NUMBER_DR	WORD
END_STRUCT	End_STRUCT

The INDEX parameter must be supplied with information before SFC51 is called. Therefore, set bit 15 of #OB82_MDL_ADDR to "1" for the event that the diagnostic interrupt was requested by an output channel. Then, program organization block OB82 as shown in figure 7.14.

```
      L         #OB82_IO_FLAG        //Query type of module
      L         B#16#54              //Load ID for input module
      ==I                            //Input ?
      JC go                          //If input, then bit 15 remains unchanged
//
      L         #OB82_MDL_ADDR       //Address supplied from local data
      L         W#16#8000            //Load hexadecimal 8000
      OW                             //"OR Word" → set bit 15
      T         #OB82_MDL_ADDR       //Save in local data
//
//****************************************************************
//
//Determine whether event is coming or going
go: L           #OB82_EV_CLASS       //Event class and identifiers
    L           B#16#39              //Identifier for coming event
    ==I                              //Coming event?
    JC come                          //Jump to read coming event
//****************************************************************
//Read and store diagnostic information
CALL SFC51      //Going event
    REQ         : =TRUE              //Always TRUE
    SZL_ID      : =W#16#00B3         //ID for data record
    INDEX       : =#OB82_MDL_ADDR    //Calculated address
    RET_VAL     : =MW 100            //RET_VAL in memory word 100
    BUSY        : =M 102. 0          //M 102. 0 is BUSY memory bit
    SZL_HEADER  : =#SZL_HEADER       //Save in local data structure
    DR          : =P#M 10. 0 BYTE16  //Read data starting with memory byte 10
    BEA

come: CALL SFC51                     //Coming event
    REQ         : =TRUE              //Always TRUE
    SZL_ID      : =W#16#00B3         //ID for data record 1
    INDEX       : =#OB82_MDL_ADDR,   //Calculated address
    RET_VAL     : =MW 104            //RET_VAL in memory word 104
    BUSY        : =M 102. 7          //M 102. 7 is BUSY memory bit
    SZL_HEADER  : =#SZL_HEADER       //Save in local data structure
    DR          : =P#M 20. 0 BYTE16  //Read data starting with memory byte 20
//
```

Figure 7.14   Calling SFC51 in OB82

To enter and test the program, follow the same procedure as described earlier, for system function SFC13. However, change the hardware configuration of the S7-400 station by removing the ET 200B 16DI/16DO module from the DP master branch and configuring the ET 200M/IM153-2 device as described in section 4.2.5. Connect the ET 200M module to the DP interface of the CPU 416-2DP. Then, load the modified hardware configuration and the new organization block OB82. To trigger a diagnostic interrupt, disconnect the supply voltage of the analog input plugged in the ET 200M device. The ET 200M device generates a diagnostic interrupt which is detected in OB82.

You may further analyze the diagnostic information provided by the SFC51 call while the system program is running, and make the user program react accordingly.

7.3.3 Diagnostics Using SFB54 *RALRM*

DP slaves or modules in DP slaves can generate different interrupts depending on their functionality. The diagnostic data sent in this way are already partly provided in the local data of the interrupt OB called. The complete diagnostic information can be read with SFB54 *RALRM* in the relevant interrupt OB.

If SFB54 is called in an OB whose start event is not an interrupt from the I/O, the SFB provides correspondingly less information at its outputs (see also Section 5.4.2, Table 5.37). In addition, a new instance DB must be used every time SFB54 is called in different OBs. If the data resulting from an SFB54 call are to be evaluated outside the associated interrupt OB, a separate instance DB must even be used for each OB start event.

SFB54 can be called in different modes. The mode is specified in the relevant input parameter of SFB54:

- In mode "0", the interrupt-generating DP slave or its module is output in the ID parameter, and the NEW output parameter receives the value "TRUE". All other output parameters are not overwritten.

- In mode "1", by contrast, all output parameters of SFB54 are overwritten with the relevant diagnostic data independently of the interrupt-generating component.

- In mode "2", SFB54 checks to see if the component specified in the F_ID input parameter has generated the interrupt. If yes, the output parameter NEW is overwritten with "TRUE" and all other output parameters are overwritten with the relevant data. If F_ID and the generating component are not the same, NEW receives the value "FALSE".

In the program example below (Figure 7.15), the diagnostic data in OB82 are evaluated with SFB54. The destination area must be sufficient here for the standard diagnostics (6 bytes), for the code-specific diagnostics (3 bytes for 12 slots), and for the evaluation of the device-specific diagnostics (another 7 bytes for the module status).

Additional bytes would need to be reserved for further evaluation (channel-specific diagnostics), provided the DP slave supports this function.

```
//...
//*****************************************************************
//Call SFB 54. DB54 is selected as the instance DB
//*****************************************************************
CALL SFB54, DB54
        MODE : =1                //1 = All output parameters are set
        F_ID : =                 //Address of the slot from which diagnostic data are
                                   to be fetched
        MLEN : =20               //Max. length of the diagnostic data in bytes
        NEW : =M80. 0            //Irrelevant
        STATUS: =MD90            //Function result, error message
        ID : =MD94               //Address of the slot from which an interrupt has been
                                   received
        LEN : =MW82              //Length of additional interr. info (4 bytes header
                                   +16 bytes diag. data)
        TINFO : =P#M 100. 0 BYTE 28 //Pointer to OB start and administration
                                      information: 28 bytes
        AINFO : =P#M 130. 0 BYTE 20 //Pointer to destination area, with diagnostic data
//*****************************************************************
//Structure of the stored diagnostic data:
// MB 130 to MB 133:           Header information (length, identifier, slot)
// MB 134 to MB 139:           Standard diagnostics (6 bytes)
// MB 140 to MB 142:           Code-specific diagnostics (3 bytes)
// MB 143 to MB 149:           Module status (7 bytes)
//...
//*****************************************************************
//Evaluation of the diagnostic data, example
//*****************************************************************
        A M 141. 0               //Slot 1 with error?
        JC stp1
BE
//*****************************************************************
//Evaluation for error on slot 1
//*****************************************************************
stp1: L       MB 147            // Fetch module status slot. 1 to 4
      AW      W#16#3            //Filter out slot 1
      L       W#16#2            //2-bit status 'wrong module' plugged in
      ==I
      S F 0. 1                  //Response to wrong module
      L MB 147                  //Fetch module status slot. 1 to 4
      AW      W#16#3            //Filter out slot 1
      L       W#16#1            //2-bit status 'invalid data'
      ==I
      S F 0. 2   //Response to invalid data
//*****************************************************************
//End of evaluation
//*****************************************************************
//...
```

Figure 7.15   Calling SFB54 in OB82

To enter and test the program, follow the same procedure as described in Section 7.3.2. Open OB82 and delete the previous program. Then enter the relevant program and load OB82 into the CPU connected via MPI. When a diagnostic interrupt is received, the diagnostic data are then read using SFB54.

You may further analyze the diagnostic information provided by SFB54 while the system program is still running. The user program can then analyze them and react accordingly.

7.4 Diagnosis Using the SIMATIC S7 Diagnostic Block FB125

The DP diagnostic block FB125 permits easy diagnostic evaluation of a DP master system from within the STEP 7 user program.

A more general diagnosis – the overview diagnosis – tells you which DP slaves are configured, which are actually present on the bus, which are faulty or have broken down. In addition, you can request a more detailed diagnosis which provides you with detailed information about a specific slave device.

Diagnostic block FB125 can be used for the following integrated and external DP interfaces.

- CPU 31x-2DP (6ES7 315-2AF01-0AB0 or higher)

- C7-626 DP (6ES7 626-2AG01-0AE3 or higher)

- C7-633 DP and C7-634 DP

- SINUMERIK 840D

- CPU 41x-2 DP

- CP 443-5

- IM 467 and IM 467 FO

- WIN AC

- WIN LC

For more detailed information visit the Internet server of Siemens A&D Customer Support at http://www4.ad.siemens.de → FIND → Search term: FB 125. Here you can also download diagnostic block FB 125.

7.4.1 Diagnostic Block FB 125

FB 125 detects DP slaves that are faulty or broken down and have therefore generated an interrupt. It displays detailed diagnostic information about the faulty or slaves, such as slot number or module number, module status, channel number, channel fault.

Table 7.14 Input parameters of FB 125

Name	Type	Comment
DP_MASTERSYSTEM	INT	No. of DP master system
EXTERNAL_DP_INTERFACE	BOOL	External DP interface (CP/IM)
MANUAL_MODE	BOOL	Manual mode for individual diagnosis
SINGLE_STEP_SLAVE	BOOL	One-by-one selection of all DP slaves
SINGLE_STEP_ERROR	BOOL	One-by-one selection of the errors on the DP slave
RESET	BOOL	Reset the evaluation
SINGLE_DIAG	BOOL	Diagnosis of individual DP slaves
SINGLE_DIAG_ADR	BYTE	DP slave address for individual diagnosis

Table 7.15 Output parameters of FB 125

Name	Type	Comment
ALL_DP_SLAVES_OK	BOOL	All DP slaves are okay
SUM_SLAVES_DIAG	BYTE	No. of slaves concerned
SLAVE_ADR	BYTE	DP slave address
SLAVE_STATE	BYTE	0: okay, 1: failed, 2: faulty, 3: not configured/cannot be evaluated
SLAVE_IDENT_NO	WORD	Id. Number of the DP slave
ERROR_NO	BYTE	Error number
ERROR_TYP	BYTE	1: slot diagnosis, 2: module status, 3: channel diagnosis, 4: S7 diagnosis
MODULE_NO	BYTE	Module number
MODULE_STATE	BYTE	Module state
CHANNEL_NO	BYTE	Channel number
CHANNEL_ERROR_INFO	DWORD	Channel error information (for standard and S7 slaves)
SPECIAL_ERROR_INFO	DWORD	Special error information (additional information for S7 slaves)
DIAG_OVERFLOW	BOOL	Diagnosis overflow
BUSY	BOOL	Evaluation in progress

7.5 Diagnosis Using a PROFIBUS Bus Monitor

Also known as SCOPE, the PROFIBUS bus monitor offers additional diagnostic utilities for PROFIBUS systems. A bus monitor usually consists of a PROFIBUS interface card which is installed in a PG programming unit or PC, and software with a Windows graphical user interface. A bus monitor records telegram communication on PROFIBUS by monitoring the bus. It does not occupy a PROFIBUS address on the bus.

Depending on the manufacturer of the device, a bus monitor has various functions and user interfaces. A bus monitor usually provides the following functions:

▷ Live list

▷ Filter

▷ Trigger

Live list

The *Live list* function identifies all devices connected to PROFIBUS through their PROFIBUS address. The devices and their PROFIBUS addresses are listed in a dialog box (figure 7.16).

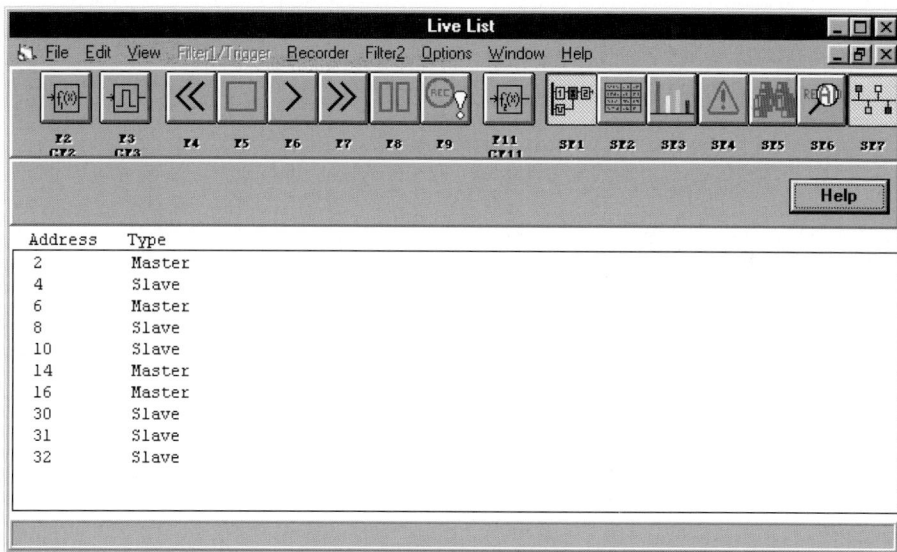

Figure 7.16   A live list

Filter

Use the *Filter* function if you want to restrict the recorded telegrams according to some defined criteria. You can usually define a second filter for the telegrams that have passed through the first filter. This allows you to narrow the list of telegrams. For example, if you define a token filter, all token telegrams are disregarded and thus not recorded.

Trigger

Use the *Trigger* function if you want to stop recording when a certain event arrives. The trigger can react to a certain PROFIBUS address or to a certain value in the data telegram, for example.

Newer bus monitors offer extended diagnostic tools. Their functional range may include:

▷ Automatic recognition of the baud rate on PROFIBUS

▷ Save telegrams in a ring buffer or in a file. Prepare and represent data for further analysis (figure 7.17)

▷ Decode telegrams and break them down further based on the selected profile (figure 7.18)

▷ Perform various statistical functions, such as counting bytes or faulty telegrams per second.

▷ Integrate a hardware trigger, for instance for triggering upon an external signal

▷ Trigger automatically upon arrival of faulty telegrams

▷ Record faulty telegrams and prepare the information for further analysis

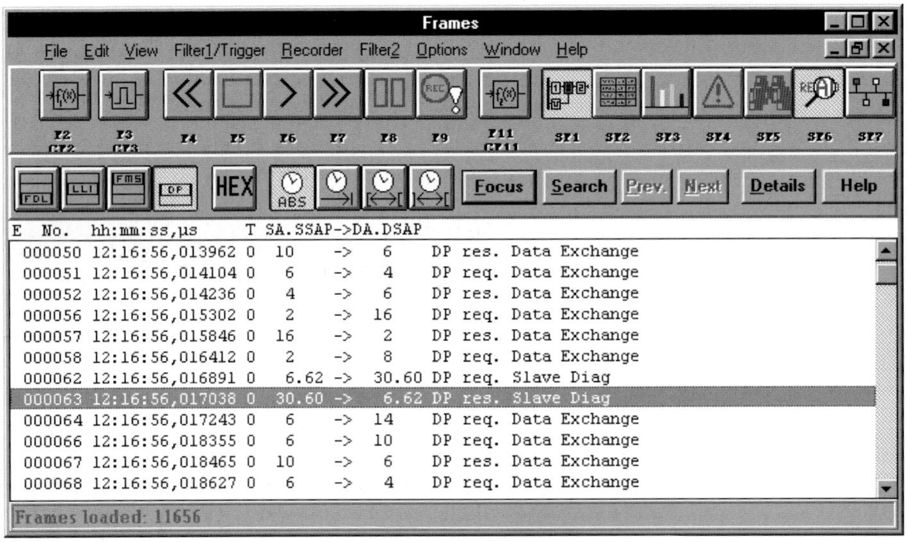

Figure 7.17 Overview presentation of the recorded telegrams

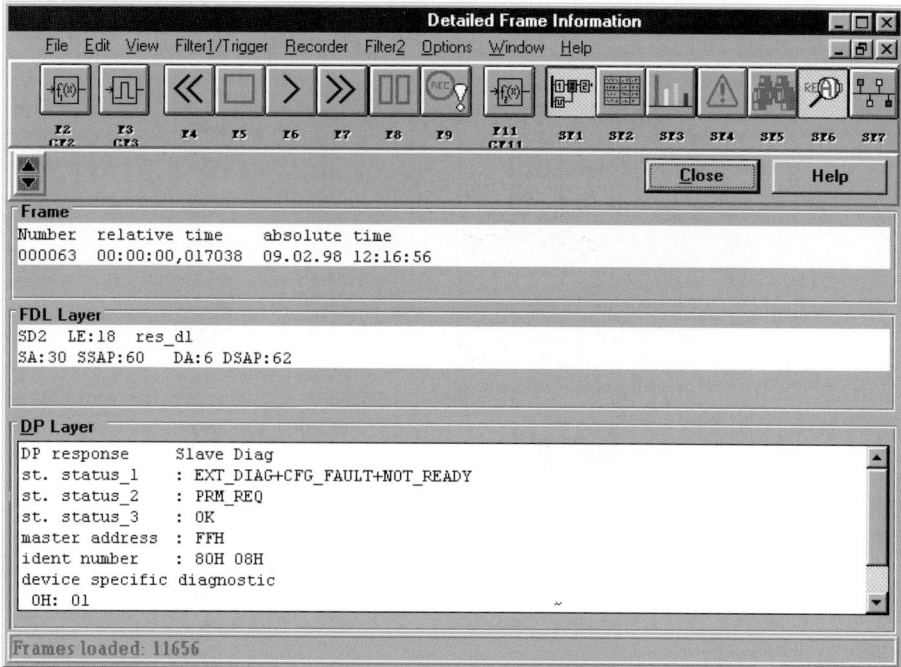

Figure 7.18   Detailed presentation of a diagnostic telegram

Even if a bus monitor offers the complete functional range just described, and the analysis of the recorded telegrams is well represented by the Windows graphical user interface, only someone with comprehensive knowledge and experience in the PROFIBUS field will be able to draw the right conclusions from the diagnostic information provided by a PROFIBUS monitor.

7.6 Diagnostics with the Diagnostic Repeater

In its basic functionality, the diagnostic repeater (Figure 7.19) is an RS 485 repeater (Section 1.4.1). However, it is also able to monitor segments of an RS 485 PROFIBUS subnet (copper cable) for physical problems during operation, and to report any cable faults to the relevant DP master.

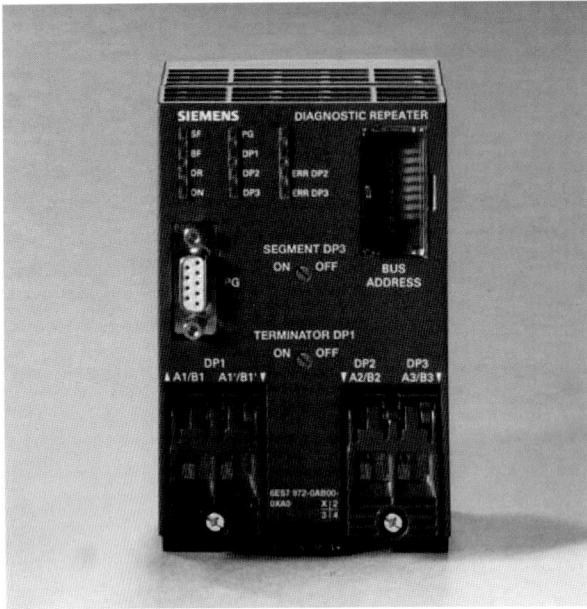

Figure 7.19   Diagnostic repeater

Along with the functions of the normal repeater such as galvanic isolation of two bus segments and the connection of more than 32 nodes, the diagnostic repeater enables the connection of a third bus segment, and permanent line diagnostics on two connected segments during plant operation.

In order to report the problems detected by line diagnostics to the DP master, the diagnostic repeater is operated as a DP slave. STEP 7 from Version 5.1 SP2 is to be used for configuring the slave functionality.

Line diagnostics of the diagnostic repeater are always carried out in two steps.

7.6.1 Determining the Topology

In the first step, the topology is determined. This is initiated once by the user. The diagnostic repeater calculates all the PROFIBUS addresses on the bus and the absolute distance of the nodes to each other. The diagnostic repeater saves the calculated values in the topology table, an internal retentive memory area, so these values will be available after any power failure.

If the physical structure of a system is changed, through adding or removing nodes, for example, the user must initiate the procedure for determining the topology again. For this purpose, the relevant project with the diagnostic repeater is opened in the *SIMATIC Manager,* the PROFIBUS object is marked, and then the function PLC/PREPARE LINE DIAGNOSTICS is selected (Figure 7.20).

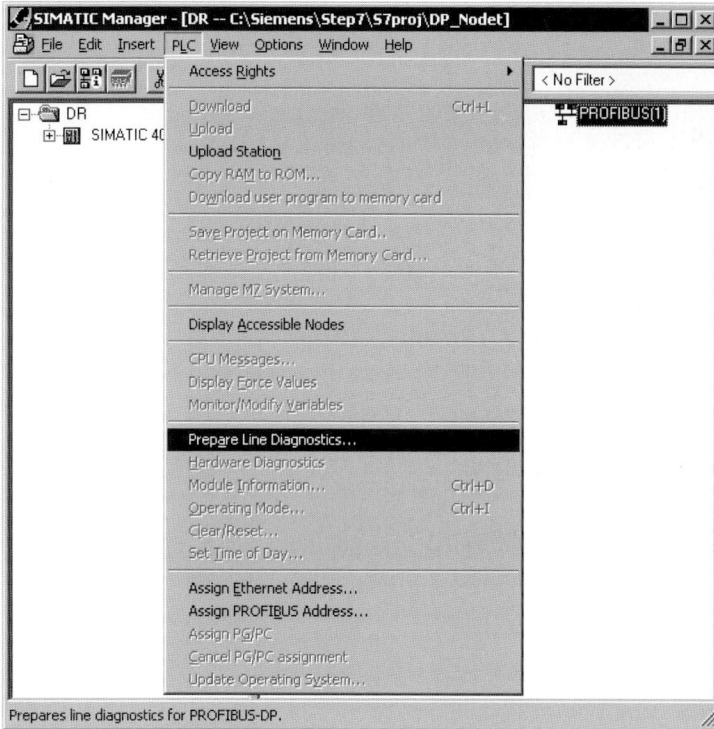

Figure 7.20   STEP 7/Prepare Line Diagnostics

7.6.2 Tracing Faults

When the topology has been determined and the diagnostic repeater and connected PROFIBUS subnets are in operation, the diagnostic repeater analyzes and evaluates the signals to the segments connected to the DP2 and DP3 connections. In addition, the distance and type of any faults are determined. When a fault occurs, the diagnostic repeater automatically transfers a message by diagnostic frame to the DP master. This message contains information on the fault location, the affected segment, and the type of fault.

The fault location is specified relative to the existing nodes on the basis of the topology table; for example, "Short-circuit of signal cable A to shield between nodes 12 and 13". However, the specified distance can have a tolerance of approximately one a meter. The fault messages are displayed graphically in STEP 7 (Figure 7.21).

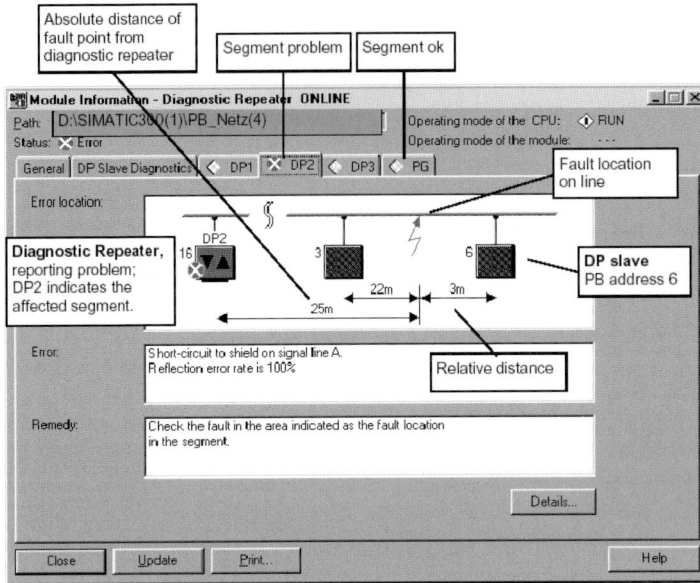

Figure 7.21   Fault message of the diagnostic repeater in STEP7/module status

The diagnostic repeater can determine the following faults:

- Cable break in signal lines A or B

- Short-circuit of the signal lines A or B to shield

- No terminating resistors

- Loose connections

- Impermissible cascade depth

- Too many nodes in a segment

- Nodes too far away from diagnostic repeater

- Faulty message frames

However, the diagnostic repeater cannot detect non-energized, additionally inserted terminating resistors. Furthermore, a short-circuit between the signal lines A and B is not detected.

7.6.3 Requirements for Operating the Diagnostic Repeater

So that the diagnostic repeater works reliably and can provide its full functionality, special installation guidelines must be followed in addition to the generally valid guidelines for installing PROFIBUS networks:

- The diagnostic repeater may not be used in purely MPI/FDL/FMS networks.

- The DP master must be operated on segment DP1.

- No spur lines are allowed in segments DP2 and DP3.

- Maximum cable length on segments DP2 and DP3 is 100 m each.

- Use of an RS 485 bus terminal is not allowed.

- The use of other components with repeater function results in the wrong topology being determined. The line is monitored only to the repeater components.

Arrangement of the diagnostic repeaters

When arranging the diagnostic repeaters, you must ensure that only one measuring circuit per segment is active. The measuring circuit is only effective on connections DP2 and DP3. If another diagnostic repeater is to be connected to these segments, it must use interface DP1. If there are two or more measuring circuits in one segment, a diagnostic message is sent to the DP master.

Figure 7.22 shows the impermissible interconnections between two diagnostic repeaters and Figure 7.23 shows the permissible interconnections.

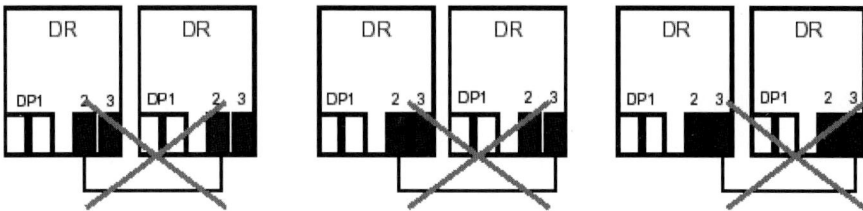

Figure 7. 22   Impermissible interconnection of diagnostic repeaters

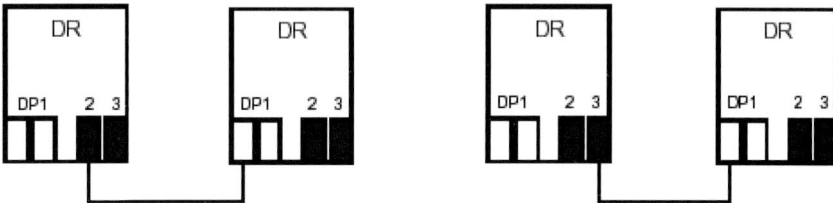

Figure 7. 23   Permissible interconnection of diagnostic repeaters

8 Setting Up and Commissioning a PROFIBUS DP System

Introduction

This chapter gives you advice on how to set up your PROFIBUS DP system using RS 485 copper cables and how to commission it and start it up for the first time. We will explain simple ways to localize and correct errors caused by incorrect cabling.

You will also learn how DP input/output signals can be tested using STEP 7 functions.

Note that the information in this chapter does not take precedence over the general guidelines on setting up electrical and electronic systems. Make sure to always comply with the guidelines of the equipment manufacturer and product-related regulations on the setup of PROFIBUS systems.

8.1 Tips on Setting Up a PROFIBUS DP System

8.1.1 System Setup with Grounded Reference Potential

The standard way to set up S7 DP systems in industrial plants is with grounded reference potential. This means that you must connect all module racks and load current circuits to a common reference potential (ground). In this way, interference currents are diverted by the connected grounding line. The bus plug connector connects the shield of the PROFI-BUS cable to all bus stations on the network. You should ensure that interference currents which may occur due to poor PROFIBUS cable layouts or plant setups are diverted as early as possible, ideally already at the housing of the control cabinet. This can be done by attaching the cable shield to the cabinet housing in such a way that the contact surface is as large as possible. Use cable clamps for this purpose.

With this type of setup, link the grounding connection of the individual components, such as the rail of the S7-300 and the ET 200M, to a common grounding point in the cabinet. This would typically be the grounding bus bar. In addition, connect the M potential (mass) of the 24 V supply voltage to the grounding point. Always make sure that the cross-sectional area of the connection lines to the grounding point is large enough. Also make sure that the individual grounding bus bars in the cabinets have the same grounding potential. This prevents potential differences and equalizing currents.

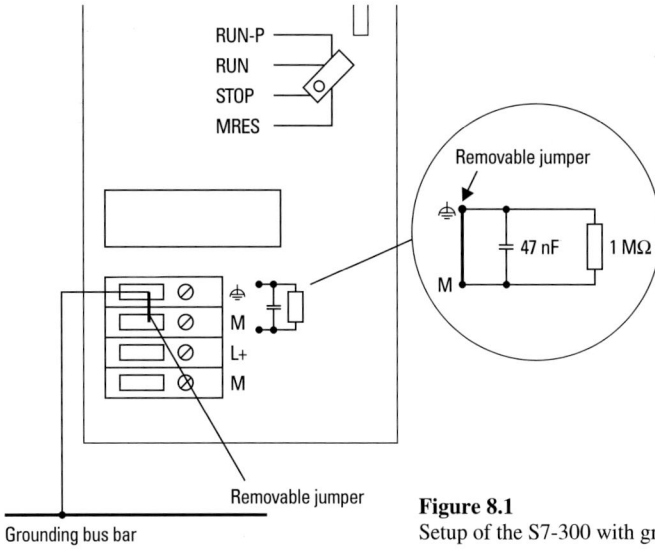

Figure 8.1
Setup of the S7-300 with grounded reference potential

S7-300 with grounded reference potential

In an S7-300 programmable controller with grounded reference potential, insert a jumper on the CPU between the M potential connection and function ground (see figure 8.1). The S7-300 CPU312 IFM module can only be operated with grounded reference potential as M potential and function ground are already connected inside the CPU.

S7-400 with grounded reference potential

In an S7-400 programmable controller with grounded reference potential, insert a jumper between the M reference potential and the connection on the module rack rail as shown in figure 8.2.

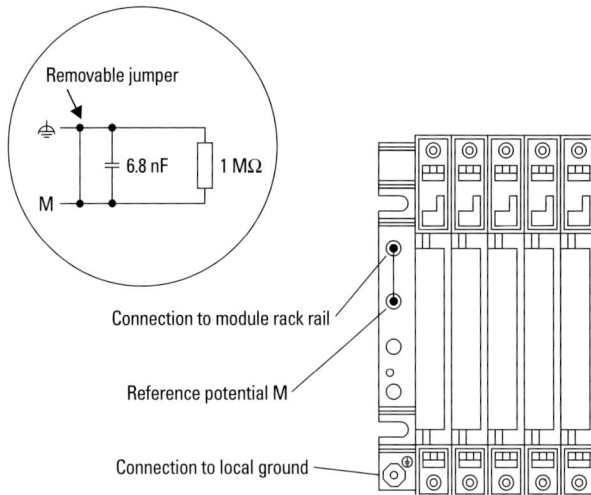

Figure 8.2
Setup of the S7-400 with grounded reference potential

The module rack itself must be connected with the grounding bus bar in the cabinet.

8.1.2 System Setup with Ungrounded Reference Potential

Some PROFIBUS network systems must be set up with ungrounded reference potential. This applies to installations that use ground fault monitoring and also to installations that extend over a large area. Such widely spread installations often produce differences in reference potential between the individual bus stations which cannot be compensated by equipotential bonding lines. In this type of setup, interference currents are diverted to ground by means of RC networks. The load voltage supplies must be free of ground potential. Similarly, the RS 485 interfaces of the connected bus stations must be floating. In this type of setup, it is important that the shield of the PROFIBUS cable only be applied on one side.

S7-300 with ungrounded reference potential

Operation of an S7-300 controller with ungrounded reference potential means that the jumper on the CPU shown in figure 8.1 between the M potential connection and function ground may not be inserted. To prevent hazardous static charging of system parts, high-frequency interference currents are diverted to ground by means of the RC network between M and function ground.

S7-400 with ungrounded reference potential

Operation of an S7-400 controller with ungrounded reference potential means that the jumper shown in figure 8.2 between M reference potential and the connection on the module rack rail may not be inserted. To prevent hazardous static charging of system parts, high-frequency interference currents are diverted to ground by means of the RC network between M and function ground.

8.1.3 Installing the PROFIBUS Cable

Due to the high electrical power required by the system, the electrical lines and cables frequently carry high voltages and currents. If such lines and cables are installed parallel to the PROFIBUS cable over long distances, capacitive and inductive interference can occur on the PROFIBUS cable which disturbs data communication in the network. To counteract this, make sure right from the beginning when the cables are installed that a distance of at least 10 cm is maintained between the PROFIBUS cable and the other power cables. Power cables and PROFIBUS cables should always be installed on separate cable and line paths.

8.1.4 Shielding the PROFIBUS Cable

Interference currents and electromagnetic interference are diverted to ground through the shield of the PROFIBUS cable. A low-impedance connection of the shield to ground potential is particularly important. The cable shield must usually be applied on both sides. Particularly in the area of higher interference frequencies, this measure provides good interference suppression. If a potential difference exists between individual bus stations of a widespread system and equipotential bonding cannot be provided, we recommend that you connect the cable shield on only one side to avoid equipotential bonding currents on the PROFIBUS cable shield. Equipotential bonding current flowing over the cable shield may reduce the effectiveness of the shield considerably.

For bus stations that are stationary we recommend that you bare the shield on the PROFI-BUS cable at the point it enters the cabinet and connect it to ground potential with appropriate cable clamps. Be careful not to damage the cable cores when stripping the cable sheath.

8.2 Tips on the Initial Start-Up of a PROFIBUS DP System

8.2.1 Bus Cables and Bus Plug Connectors

PROFIBUS cables and bus plug connectors are important components of a DP system. Errors during installation and connection of the bus cables can interfere considerably with data communication between the bus stations. Since serious errors, such as reversed data lines, line breaks or short circuits make data communication impossible, the installation of the bus cables and bus plug connectors, and correct insertion of bus terminating resistors should be checked before a PROFIBUS DP system is switched on for the first time.

8.2.2 Checking the PROFIBUS Bus Cable and the Bus Plug Connectors

Incorrect connection of the PROFIBUS cable to the bus plug connector can cause problems with data communication. You can use the following simple test to detect such fundamental errors and correct them.

Figure 8.3 shows the principle of a test which detects reversed data lines on the standard bus plug connectors. This test requires that the bus plug connectors mounted on the bus cable are not connected with the bus stations. There must also not be any bus terminating resistors on the bus plug connectors of the bus line.

To perform the measurement(s), you will need two single 9-pin sub D plug connectors with socket contact. Auxiliary plug connector 1 contains a 1-pin change-over switch whose floor contact is connected with the shield (housing) of the 9-pin sub D plug connector with socket contact. The two circuit contacts are connected with pin 3 (data line B) and pin 8 (data line A).

217

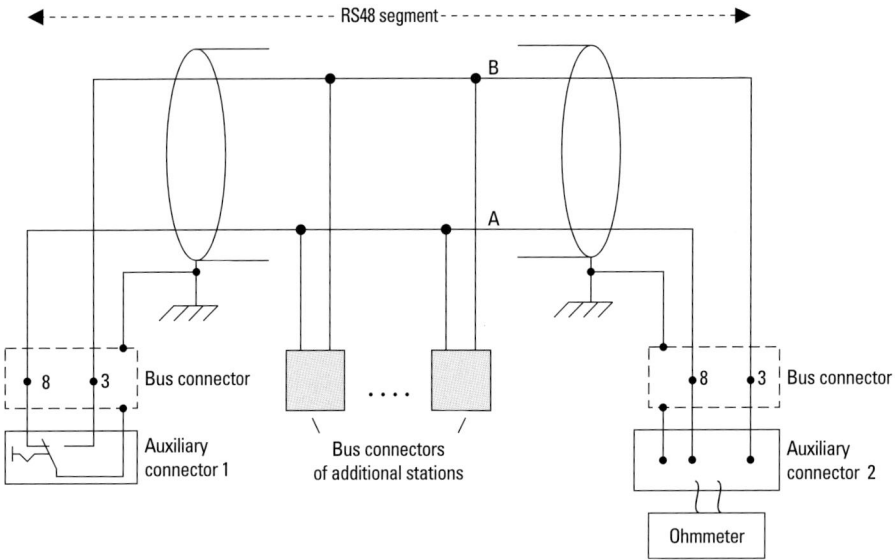

Figure 8.3   Block diagram for testing the PROFIBUS cable

Auxiliary plug connector 2 is a simple adapter (connection sockets) to connect an ohm meter to the bus plug connector.

To test the data line, plug the two auxiliary plug connectors 1 and 2 on the bus plug connectors at the beginning and end of the segment of the bus line. You can check the following points on the bus line by measuring the resistance at contacts 3 and 8 and on the shield of auxiliary plug connector 2, while setting the switch on auxiliary plug connector 1 accordingly.

▷ Simple wire mix-up on the data lines

▷ Interruption of one of the two data lines

▷ Interruption of the line shield

▷ Short circuit between the data lines

▷ Short circuit between the data lines and the cable shield

▷ Too many bus terminating resistors inserted (unintentional)

Before you evaluate the results of these measurements, take into account the type of line used (see table 1.2) and the loop resistance of the bus line which varies with the line length.

This test allows you to find errors without opening the bus plug connector by moving from one bus plug connector to the next bus plug connector and thus approaching auxiliary plug connector 1 in the direction of auxiliary plug connector 2. On auxiliary plug connector 2, carry out a control measurement with each change in the plug connection.

To test the data line, follow the procedure described below.

The measurements are listed below and shown in figures 8.4 to 8.6, starting with the switch position on auxiliary plug connector 1 and the connection of the measuring instrument on auxiliary plug connector 2 (configurations A to D).

Performing the measurements

Configuration A:

On auxiliary plug connector 1, set the switch to position 3 (connection of pin 3 with the shield). Connect the measuring instrument to auxiliary plug connector 2, to pin 3 and to the shield.

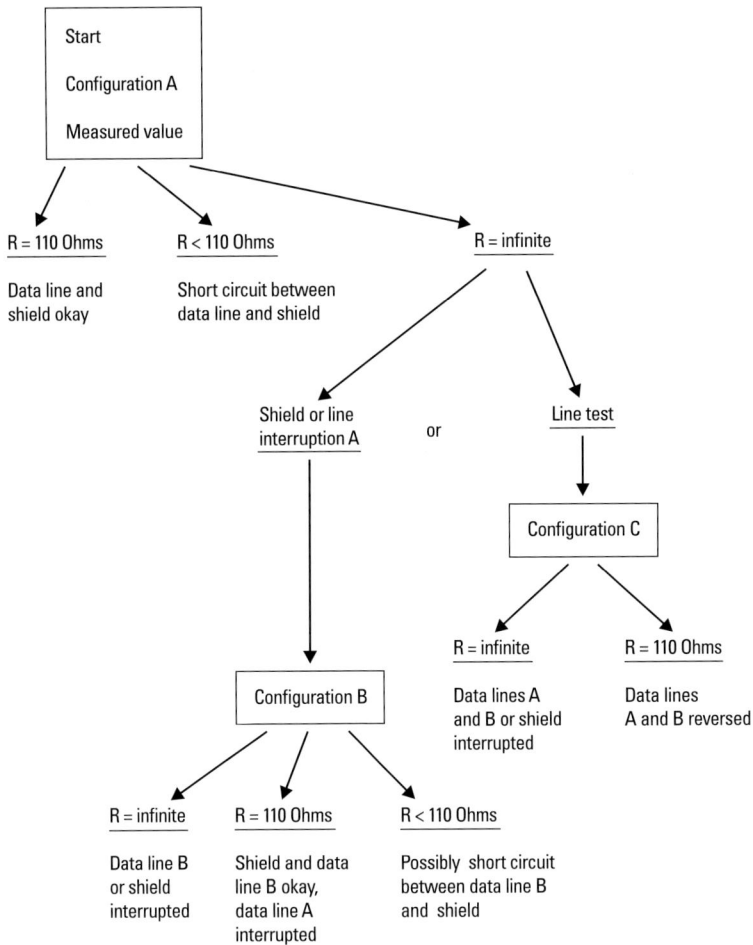

Figure 8.4   Checking the bus line, part 1

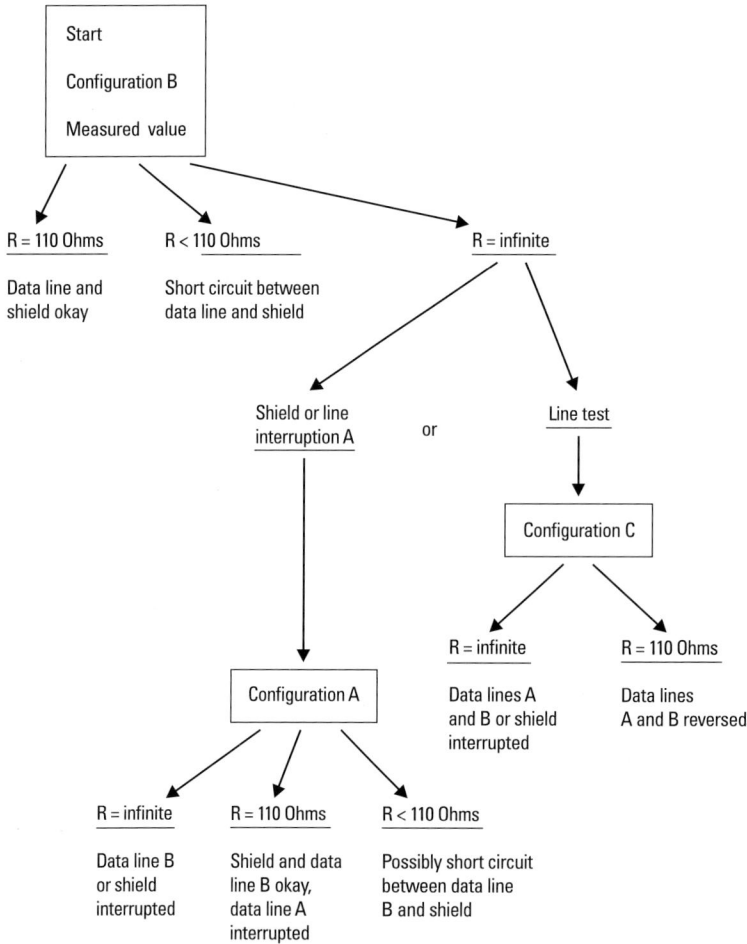

Figure 8.5 Checking the bus line, part 2

Configuration B:

On auxiliary plug connector 1, set the switch to position 8 (connection of pin 8 with the shield). Connect the measuring instrument to auxiliary plug connector 2, to pin 8 and to the shield.

Configuration C:

On auxiliary plug connector 1, set the switch to position 3 (connection of pin 3 with the shield). Connect the measuring instrument to auxiliary plug connector 2, to pin 8 and to the shield.

Configuration D:

The switch position of auxiliary plug connector 1 is irrelevant. Connect the measuring instrument to auxiliary plug connector 2, to pin 3 and to pin 8.

```
┌─────────────────┐
│  Start          │
│                 │
│  Configuration D│
│                 │
│  Measured value │
└─────────────────┘
      ↙        ↘
```

R = 110 Ohms + x Ohms   R < 110 Ohms + x Ohms

Terminating resistors   Too many terminating
correct.   resistors inserted
Everything okay.

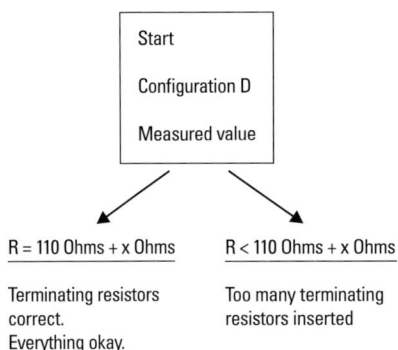

Figure 8.6   Checking the bus terminating resistors

8.2.3 Bus Termination

The active bus termination, consisting of a combination of termination resistors (see also figure 1.6), prevents reflections during data transmission and ensures a defined idle state potential on the data line when no station is active on the bus. There must be an active bus termination at the beginning and end of the RS 485 bus segment.

Lack of bus termination will cause interference during data communication. Too many combinations of termination resistors will also cause problems since each bus termination also represents an electrical load and the transmission level required for a high signal-to-noise ratio can no longer be ensured for data transmission. Too many or two few bus terminations can also cause sporadic transmission interference. This applies in particular when a bus segment is operated at the electrical power limit determined by the maximum number of bus stations, the maximum bus segment length and the maximum transmission speed which can be selected.

The voltage supply required for the active bus termination is usually taken directly from the bus plug connector of the connected bus station. If it is obvious at the outset, when planning a system, that the voltage supply required for the active bus termination during system operation cannot be ensured, then appropriate measures must be taken. A typical example is the situation in which the bus station powering the bus terminating resistor is frequently isolated from power or disconnected from the bus. In this case, use a bus termination with external power supply or a repeater for the affected bus termination of the bus station.

8.3 BT 200 Test Device for the Bus Physics of PROFIBUS DP

The BT 200 test device shown in figure 8.7 is an easy-to-use hand-held device which offers various functions for diagnosing PROFIBUS DP bus systems without additional measuring aids such as PC/PG or oscilloscope.

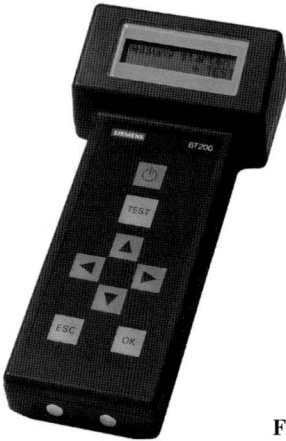

Figure 8.7   Test device BT 200 for PROFIBUS DP

8.3.1 Wiring Test

To test the wiring of a bus segment, test the line between the BT 200 test device and the test plug connector. Plug-to-plug testing can be performed during the installation phase. Always insert the test plug connector at one end of the bus segment. This wiring test also detects short circuits outside the testing path (see figure 8.8). Make sure that only the beginning and the end of the bus segment are fitted with terminating resistors.

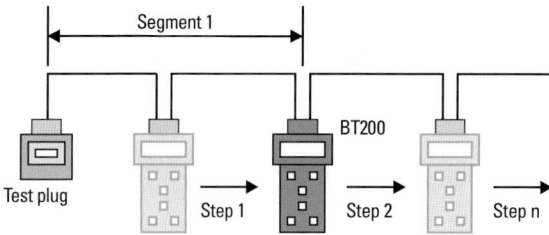

Figure 8.8   Testing the bus wiring with the BT 200 test device

8.3.2 Station Test (RS 485)

You can apply the test illustrated in figure 8.9 to check the RS 485 interface of an individual DP slave. The test device checks the RS 485 drivers, the supply voltage, and the RTS signal.

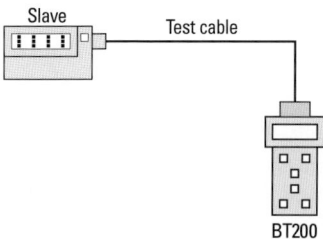

Figure 8.9   Testing the RS 485 interface with the BT 200 test device

8.3.3 Branch Test

You can use the BT 200 test device to check the availability of all slaves that are connected to the PROFIBUS network. It also allows you to address an individual slave which verifies the stations' bus address settings at the same time. You can extend the branch test across repeaters/fiber optic conductors.

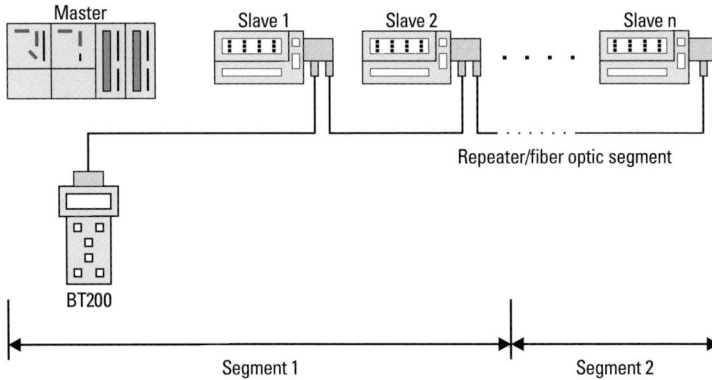

Figure 8.10 PROFIBUS DP branch test with the BT 200 test device

8.3.4 Distance Measurement

Use distance measurement to measure the length of the installed PROFIBUS cable. In this way you can verify that the maximum permissible bus segment length has not been exceeded.

8.3.5 Reflection Test

The reflection test determines points of interference, such as short circuits or line interruption. The test indicates the distance between the BT 200 test device and the position of the fault. You may even carry out this test on stations connected to the bus and their power supplies switched on. Note though that there must be no data exchange while you are executing the reflection test. This means that the DP master must be switched off or not connected to the bus.

Reflections can occur under the following conditions.

• Short circuit

• Wire break

• Stub lines (too many or too few) are used.

• Too many or no terminating resistors are inserted.

• An unsuitable cable type is used within the measuring path (PROFIBUS line).

• Cables are not installed correctly (non-permissible terminal connections).

8.4 Signal Test of the DP Inputs and Outputs

Commissioning a DP system should also include a thorough wiring test of the signal paths of the sensors and actuators connected to the DP slaves. To test the signals of the DP inputs and outputs, use the STEP 7 function *Monitor/Modify Variables*.

To perform the signal test, the CPU must be in the STOP state. To stop the CPU, you can use the operating mode switch of the CPU or the STEP 7 function PLC → OPERATING STATE in the online view of *SIMATIC Manager*.

Connect the PG programming unit or PC to the CPU using the MPI cable. Select ACCESSIBLE NODES and *MPI="Address"* in the shortcut menu and PLC → MONITOR/MODIFY VARIABLES.

Enter the DP input/output bytes to be tested. Use the direct I/O addresses - PIB/PIW/PID for inputs and PQB/PQW/PQD for outputs.

As shown in figure 8.11, select VARIABLE → ENABLE PERIPHERAL OUTPUTS to open the dialog box for enabling the outputs. Answer YES to activate the "enable peripheral outputs" mode. This disables the OD signal (*Output-D*isable) of the CPU. This signal prevents output modules from issuing values when the CPU is in the STOP state.

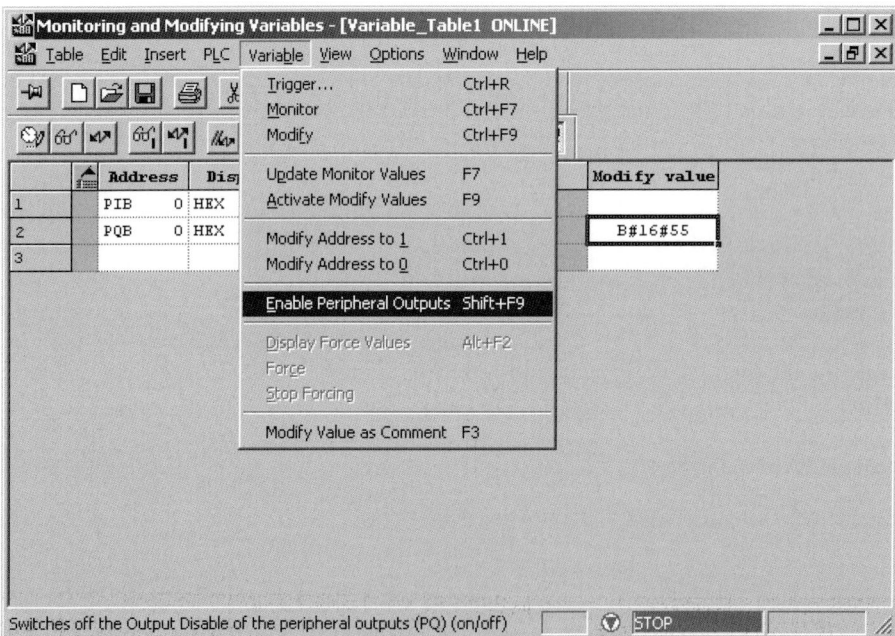

Figure 8.11   Enable peripheral outputs (PQ)

Now, enter the required "modify value" for the outputs you want to test. Use ACTIVATE MODIFY VALUES to connect the specified "modify value" to the defined output address. The ACTIVATE MODIFY VALUES function is not cyclic. You must reactivate it for each newly specified "modify value" that is to be connected to the output. To check the input states, use the function UPDATE MONITOR VALUES (see figure 8.12).

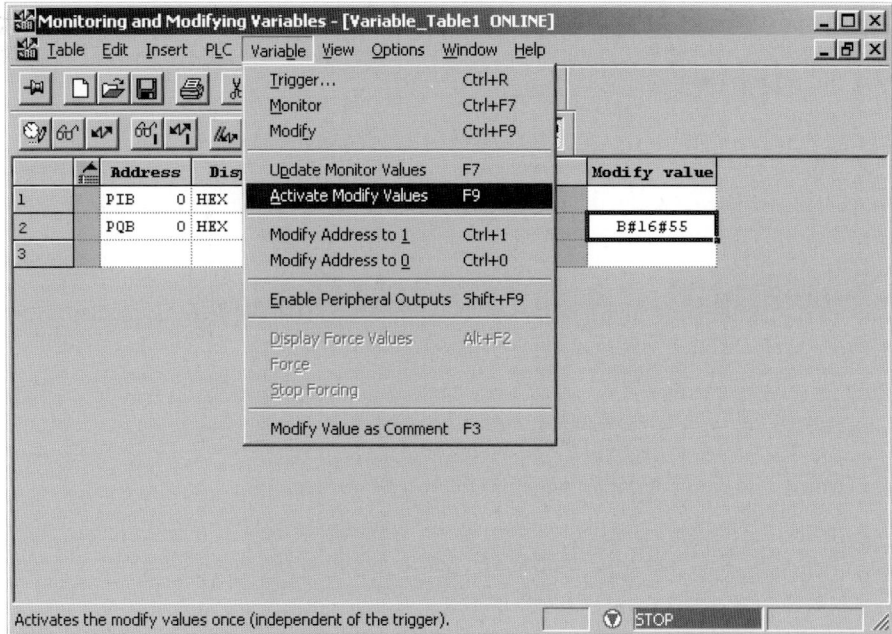

Figure 8.12 Signal test of the DP inputs/outputs using *Monitor/Modify Variables*

9 Other DP-Related STEP 7 Functions

9.1 GSD Files

The GSD files for DP slaves and Class 1 DP masters contain characteristic device properties of these DP components. GSD files have standardized characteristics, such as predefined "DP key words" and fixed file formats (syntax). You can therefore edit PROFIBUS standard GSD files using a non-proprietary configuration tool.

GSD files allow you to already check PROFIBUS device data for plausibility, validity and correct performance in the early configuration phase. This helps you to prevent errors which you would otherwise only detect when the DP device is connected and running.

GSD files are ASCII files. You can create and edit them using any ACSII text editor. For the standardized key words and required setup of the GSD files, see volume 2 of the EN 50 170 standard. The names of GSD files relate to the manufacturer of the DP device and the device designation, and must comply with the PROFIBUS DP name convention for GSD files.

The PROFIBUS International PNO provides a GSD editor on its Internet server. Visit http://www.profibus.com to download this editor. You can use the GSD editor to create new GSD files and check already existing ones.

9.1.1 Installing a New GSD File

To install a new GSD file, open the hardware configuration tool *HW Config*. In the menu bar, select OPTIONS → INSTALL NEW *.GSE FILES .... (Yes, GSE here at this point really means GSD file!). You will always have to install a new GSD file when you want to add a new DP device to your PROFIBUS DP system configuration and the configuration tool you are using does not recognize this device.

Store the newly created GSD file under STEP 7 in the ...\Siemens\Step7\S7data\Gsd folder, and the related pictogram as bitmap file in the ...\Siemens\Step7\S7data \Nsbmp folder.

9.1.2 Importing a Station GSD File

STEP 7 saves all GSD files of the DP devices of a PROFIBUS DP system configuration in the project. This gives you freedom from the STEP 7 configuration tool as you can always move this STEP7 project to another tool and process it there, even if this configuration tool does not possess the GSD files for the new DP devices.

GSD files which are only stored in the PROFIBUS DP hardware project and not in the general STEP 7 GSD directory can only be applied to this particular project. Therefore we recommend that you import these GSD files into the general GSD folder under STEP 7 so that you can reuse them for other projects. In the menu bar of the *HW Config* program, select OPTIONS → IMPORT STATION *.GSE FILES... and store the new GSD files in the ...\Siemens\Step7\S7data\Gsd folder.

9.2 Assigning and Changing the PROFIBUS Address

Some types of DP slaves do not provide hardware switches for setting the PROFIBUS address. Instead, their bus station address is assigned by means of the DP-master class-2 function *Set_ Slave_Add*. Due to its integrated MPI online interface, the STEP 7 configuration software is capable of processing this addressing function. Note that this method of address assignment only applies to those DP slave devices that support the *Set_Slave_Add* function (see also section 2.1.3). The ET200C electronic terminal from Siemens, for instance, is such a device.

To assign the bus station address to a DP slave device using the *Set_Slave_Add* function, open *SIMATIC Manager* or *HW Config*. In the menu bar, select PLC → ASSIGN PROFIBUS ADDRESS... (see figure 9.1). Remember that the DP slave must be connected to the MPI interface of your PG programming unit or PC with an appropriate PROFIBUS or MPI cable. When you start the *Assign PROFIBUS Address* function, STEP 7 searches for the address of the connected DP slave and displays it in the "Assign PROFIBUS Address" dialog box, in the "Current PROFIBUS address" field (see figure 9.1).

The slave device's default address set by the manufacturer is 126. In the European EN 50 170 standard, this address has been reserved and cannot be used by PROFIBUS DP users. It indicates that the DP device supports the DP-master class-2 function *Set_-Slave_Add*. Note though, that you will see the default address 126 only if the slave device

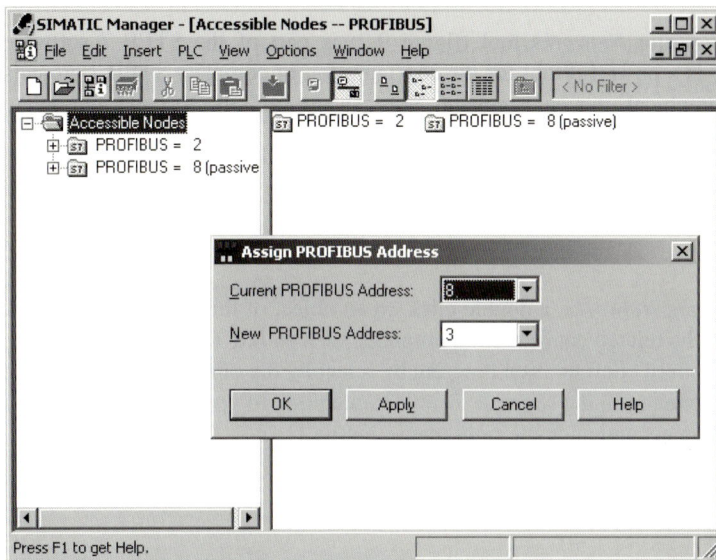

Figure 9.1 STEP 7 function *Assign PROFIBUS Address*

▷ Live list of all devices connected to PROFIBUS

▷ Read PROFIBUS statistics

▷ General diagnosis of the DP master (figure 9.3)

▷ Individual diagnosis of the DP slaves communicating with this DP master (figure 9.4)

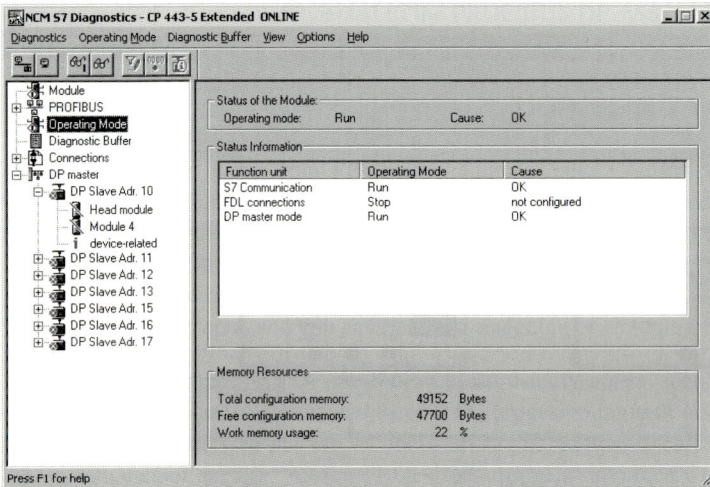

Figure 9.3   General diagnosis of the DP master in *NCM Diagnostics*

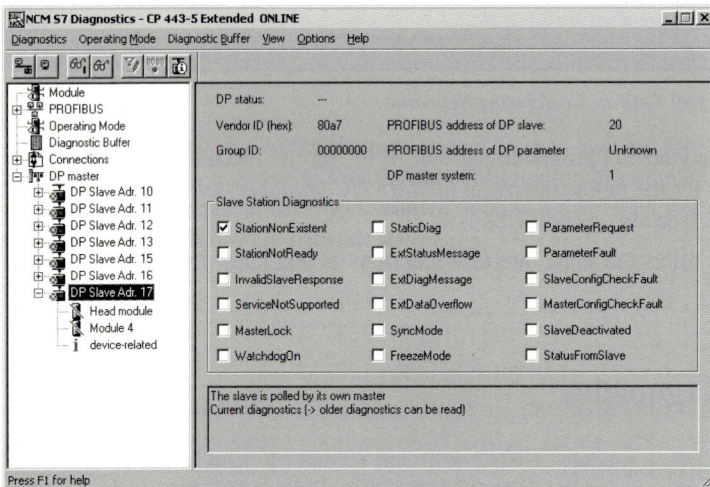

Figure 9.4   Diagnosis of individual DP slaves in *NCM Diagnostics*

Glossary

Actual parameter
Actual parameters replace formal parameters when a function block (FB) or a function (FC) is called in the user program. Example: the formal parameter "REQ" is replaced by the actual parameter "I 3.6."

Address
Identifies an operand or operand area. Examples: input I12.1; memory word MW25; data block DB3.

Addressing
Assigning an address in the user program. Addresses are assigned to operands or operand areas (e.g., input I 12.1; memory word MW25) with the aim to identify their memory location unambiguously.

Baud rate
Data transmission speed. The baud rate is the number of bits transmitted in one second (baud rate = bit rate). PROFIBUS DP allows baud rates in the range from 9.6 kBaud to 12 MBaud.

Bus
Common transmission path (transmission medium) that connects the nodes or stations to the network. In a PROFIBUS network, the bus is a twisted-pair or optical fiber cable.

Bus plug connector
Physical connection between the station (also called "node") and the bus line. In a PROFI-BUS network, bus plug connectors can be with or without a connection to the PG programming device, and are available in protection class IP 20 and IP 65.

Bus segment
Due to the physics of the network, a PROFI-BUS network can only be built to its maxi-mum size and maximum number of connected stations, if it is divided into segments that are connected with each other through repeaters.

Bus system
All stations that are physically interconnected through a bus cable, form a bus system.

Chassis ground
The entirety of all inactive parts of electric plant components which even in the events of faults do not conduct shock-hazard voltages.

Class 1 Master
The DP master device that handles the exchange of user information.

Class 2 Master
The DP master device that handles functions of network control, commissioning and configuration.

CLEAR
Operating mode of the DP Master. In this mode, the DP master reads the input data cyclically, whereas the outputs remain set at "0".

Clear/Reset
Clearing or resetting the CPU of a SIMATIC S7 programmable controller resets the CPU's main memory, the write/read area of the load memory, the system memory. The MPI parameters and the diagnostic buffer are not deleted.

Client/server principle
Data exchange according to the client/server principle implies that it is always the client station that sends a request for communication. The server responds to the request.

Combi-master
Master device that can be used as a DP master and as an FMS master.

Communication relationship
In a PROFIBUS FMS network, the communication relationship describes the logical interaction between two bus stations.

Communications protocol
A set of rules or standards designed to enable computers to connect with one another and to exchange information with as little error as possible. The communications protocol defines different aspects of communication, such as the data format of the information to be exchanged and the data flow during transmission.

Configuration
In relation to automated plants, the entire interconnected set of hardware and software components, containing their descriptions, characteristics and response definitions.

Configuring
Procedure of selecting and assembling individual hardware and software components of an automation system and adjusting their characteristics and response definitions to the automation task in question. Example: adjusting the hardware by setting the parameters of the plug-in modules of a programmable controller.

Consistent data
Input and output data area that is contiguous and cannot be separated. Consistent data cannot be stored in a byte or word structure and must therefore always be handled as an entity.

Constant
Descriptive names for constant values used by program codes. The use of constants makes it easier to read and understand the purpose of a program code. Example: A function block FB has the parameter "Max-Loops." During program execution, when the block is called in the user program, "Max-Loops" is replaced by the value declared for this constant (e.g., 10).

Constant bus cycle
The bus cycle of a PROFIBIS subnet is said to be "constant" or "equidistant" when the time interval between successive send authorizations for the DP master is constant. You define in *HW Config* whether or not the bus cycle for the PROFIBUS subnet is to be constant.

Control command
Special commands sent by the DP master to a group of DP slaves in order to synchronize input and output data. The control commands FREEZE and SYNC allow event-controlled synchronization of DP slaves.

Control command FREEZE
see FREEZE

Control command SYNC
see SYNC

Cross communication
also called "Direct communication"
In cross communication mode, the DP slave does not use a one-to-one telegram to reply within its allocated response period. Instead, it uses a special one-to-many telegram. The effect of cross communication is that the input data of the DP slaves is available to all other DP stations connected to the bus.

Cycle time
The time required by the CPU to process the user program once, from the first to the last statement.

Cyclical processing
With cyclical processing, the DP master addresses the DP slaves continuously at regular intervals. In doing so, the DP master reads the input data from the slaves and transmits output data to the slaves.

Data block (DB)
Data area in the user program that contains user data. In the SIMATIC S7 series, there are global data blocks and instance data blocks. Global data blocks can be accessed from any point in the user program. Instance data blocks are allocated to the call of a specific function block (FB).

Diagnostic buffer
Battery backed memory area in the CPU which stores all diagnostic events in the order of their occurrence.

Diagnostic data
Detailed information about the cause and location of the fault (e.g., diagnostic event). This information is part of the error message sent to the CPU.

Diagnostic interrupt
Modules that support diagnostic functions recognize system errors and report these to the CPU by generating and transmitting diagnostic interrupts.

Diagnostics
Recognition, localization, indication and evaluation of faults and error signals. The SIMATIC S7 diagnostic utilities provide monitoring functions that are executed automatically during online plant operation. This results in a higher degree of availability of the automated plant.

Direct communication
see "Cross communication"

Distributed I/O devices
Input and output modules that are not plugged in the central rack or expansion rack, but are installed at a distance from the CPU. Modules such as ET 200B, ET 200C, ET 200L, ET 200M, ET 200U, S5-95U, DP/AS-I Link and S7-300 stations with PROFIBUS DP interface can be used as distributed I/O modules.

Downloading to programmable modules
Copying objects of a configuration such as blocks of program code from the programming device to the load memory of a programmable hardware module, either directly from the programming device that is connected to the CPU, or indirectly from a remote programming device that is connected to the PROFIBUS network.

DP
*D*istributed *P*eripherals. These are input and output modules that are installed at a distance from the CPU. The connection between the

PLC and the distributed peripheral equipment is made through the PROFIBUS DP network.

DP Master
Master device which uses the PROFIBUS DP communications protocol and whose response complies with the EN 50 170 standard, Volume 2, PROFIBUS.

DP Slave
Slave device which uses the PROFIBUS DP communications protocol and whose response complies with the EN 50 170 standard, Volume 2, PROFIBUS.

DP standard
Bus communications protocol for DP (*D*istributed *P*eripherals) in accordance with EN 50 170, Volume 2, PROFIBUS.

DPV1 slave
A DP slave in accordance with EN 50170, Volume 2 or IEC 61158-3 with specific expansions of the interrupt model and of the acyclic data traffic.

Equidistant DP cycle
see "Constant bus cycle"

Error, asynchronous
Runtime error which cannot be related to a specific point in the user program. This could be for example a power supply fault or an exceeded CPU cycle. The operating system reacts to this type of error by calling up the relevant error organization block (OB). The OB is programmed by the user and contains instructions on how to react to the error.

Error OB
Organization blocks (OBs) that are reserved for error handling. An error OB contains a user program that tells the CPU how to react to a specific type of error. However, a programmed reaction to errors is only possible if the error did not cause the PLC to go into the STOP state. Different error OBs are called for different types of errors. There is for example a specific error OB for addressing errors, one for access errors in S7, etc.

Error, synchronous
Runtime error which can be related to a specific point in the user program. This could be for example an unsuccessful attempt of addressing an I/O module. The operating system reacts to this type of error by calling up the relevant error organization block (OB). The OB is programmed by the user and contains instructions on how to react to the error.

Errors, reaction to
see "reaction to errors"

FDL
Fieldbus Data Link. Layer 2 of the ISO Reference Model as it is used in a PROFIBUS network. Layer 2 consists of Fieldbus Link Control (FCL) and Medium Access Control (MAC).

Field device
Programmable controller or distributed automation device installed in the field of an automated plant, that is, in the vicinity of the machines and sensors and actuators.

Fieldbus
Digital, serial data network for multipoint communication. Widely used as process control local area network defined by the ISA S50.02 standard.

Floating
On modules with floating inputs and outputs, the reference potentials of the control and load circuits are electrically isolated from each other. The input and output circuits have no common root, that is, they have no common reference potential (so-called 1-root). Not to be confused with "optically isolated."

FM
Function Module. Hardware module that processes signals and information coming from the plant and thus takes the load of the CPU of the S7-300 and S7-400 programmable controllers. A function module is dedicated to a specific task such as counting, positioning, closed-loop control etc., which would use too much time and memory resources if

executed by the CPU. Function modules usually use the internal communication bus to exchange data with the CPU.

FMS
Fieldbus Message Specification. With PROFIBUS, Layer 7 of the ISO Reference Model. The FMS contains the protocol engine, generates the PDUs, and codes, decodes and interprets the protocol data unit.

FMS (communications) protocol
Protocol that manages the exchange of data through a PROFIBUS network in accordance with the Fieldbus Message Specification.

FMS connection
Communication link between two FMS stations.

FMS master
Master device which uses the FMS communications protocol and whose response and characteristics comply with the EN 50 170 standard, Volume 2, PROFIBUS.

FMS service
FMS services organize the data exchange between FMS stations (also called "nodes"). We distinguish between acknowledged and non-acknowledged FMS services. With acknowledged FMS services such as MSAZ, the slave returns an acknowledge telegram to the master to confirm the receipt of the FMS service. With unacknowledged FMS services such as broadcasting, the master receives no confirmation.

FMS slave
Slave device which uses the FMS communications protocol and whose response and characteristics comply with the EN 50 170 standard, Volume 2, PROFIBUS.

FMS station
(also called "FMS node")
FMS master or FMS slave connected to a PROFIBUS FMS network.

Formal parameter
Dummy parameter of an FB/FC/SFB/SFC software block which is replaced by the actual parameter when the block is called in the user program. In software blocks of type FB

and FC, the formal parameters must be declared by the user. In SFB and SFC blocks, they are present by default. When the block is called, the formal parameter is replaced by the actual parameter and the block uses the actual value provided by this parameter for program execution. Formal parameters are part of the software block's "local data" and are divided into output parameters, input parameters and input/output parameters.

FREEZE

Control command sent by a DP master to group of DP slaves. Upon arrival of the FREEZE command, the DP slave freezes the actual state of the inputs and transmits these cyclically to the DP master. With each newly arriving FREEZE command, the DP slave freezes the input values anew. Cyclical transmission of input data from the DP slave to the DP master is not resumed until the DP master sends an UNFREEZE command.

FREEZE, control command

Control command sent by the DP master to a group of DP slaves. Upon the arrival of a FREEZE command, the DP slaves maintain the states of their inputs stationary until they receive a canceling command.

GAP area

GAP update factor. The distance from the own PROFIBUS address of a master to the PROFIBUS address of the next master is called "gap." The gap update factor defines how many times the token must circulate in the network before a master checks whether there is another master in the gap. For example, a gap update factor of 3 means that the token circulates about three times before every master in the network checks whether there is a new master between its own PROFIBUS address and the PROFIBUS address of the next master.

GAP factor

Defines how many times the bus is circulated before the master searches for new active stations in order to accept them in the network and pass the token to these stations. The area between the own station address and the next

address is the GAP area. Exception: the area between the highest station address and the address 127 is not part of the GAP area.

Ground, connecting to

Deliberately connecting a conducting part of the plant to earth, for safety reasons.

Ground-free installation

Installing electrical equipment without establishing a connection to earth. In most cases, RC networks are used to suppress interference.

Group

When control commands such as FREEZE or SYNC must be transmitted to DP slaves, the DP slaves are organized into groups that can be addressed by these commands. A group can contain several DP slaves. A DP slave can be member of more than one such DP slave group, but can only belong to one master system.

Group error

Common error indicated by an LED on the front plate of a module (applies to S7-300 only). The LED indicates any type of fault that occurs on the module (internal error and external error).

GSD file

Contains the characteristic data (*Geräte Stamm Daten*) of a PROFIBUS DP device. The GSD file is usually supplied by the manufacturer of the device on a diskette and can be considered as an electronic data sheet. You need the GSD file when you want to configure the device as a station in a PROFIBUS DP network.

Hardware interrupt

see "Process interrupt"

HOLD, operating mode

The PLC goes to the HOLD state if an appropriate request is initiated on the PG programming device while the CPU is in the RUN mode. Some tests are still possible, even with the CPU being in the HOLD state.

Hot restart

When the CPU is started by moving its mode selector switch from STOP to RUN, or by switching on its power supply, the startup organization block OB 101 (hot restart, for S7-400 systems only)) or OB 100 (warm restart) is executed before cyclical program execution starts. The hot restart procedure first reads the process-image input table and then executes the cyclical organization block OB1, starting from the point of interruption due to a STOP or power failure.

Ident number
A 16-bit number that identifies a PROFIBUS product. This number is issued by the PROFIBUS User Organization. It relates the product to the associated GSD file. PROFIBUS devices which are of modular design, or which are part of a series of related products and therefore described by a single GSD file, are often identified by one and the same ident number for the entire product series.

Input parameters
Only functions (FC) and function blocks (FB) make use of input parameters. Input parameters provide the called software block with the data that it requires for correct program execution.

Insert/remove interrupt
An insert/remove interrupt is triggered by the removal or insertion of a module in a DPS7 or DPV1 slave.

Installation
also called "Plant"
The whole of the electrical resources making up a usually complex installation in manufacturing or process engineering industries. The industrial plant usually comprises, among other things, programmable logic controllers (PLCs), devices for operator control and process monitoring, bus systems, field devices, drives, supply cables.

Instance DB
An instance data block stores the formal parameters and static data of function blocks and system function blocks. An instance DB is required for an FB/SFB call.

Insulation monitoring
Monitoring of the insulation resistance of an installation or plant.

Interrupt
SIMATIC S7 distinguishes between 10 different priorities according to which the user program, or parts of it, are executed. Interrupts, such as process or hardware interrupts, belong to the criteria that determine the priority of program execution. Upon generation of an interrupt, the operating system automatically calls and executes the organization block (OB) that is related to that type of interrupt. The user program contained in this OB defines the reaction to the interrupt.

ISA
Acronym for *Industry Standard Architecture*. A bus design specification that allows components to be added as cards plugged into standard expansion slots in IBM PCs and compatibles. The ISA bus is an expansion bus for XT and AT computers (standardized 16-bit data bus and 24-bit address bus).

ISO
Short for *International Organization of Standardization*. An international association of countries, each of which is represented by its leading standard-setting organization – for example, ANSI for the United States. The ISO works to establish global standard for communications and information exchange. Its headquarters are in Geneva, Switzerland.

Loading into the PLC
Loading of loadable objects (e.g. code blocks) from the programming device into the load memory of a connected programmable module (e.g. CPU). This can be done using a programming device connected to the programmable module direct or via PROFIBUS, for example.

Local ground
Permanent connection to local ground by means of a low-ohmic connection of sufficient conductivity to divert overvoltages from electrical equipment or persons (in accordance with the DIN EN 61 158-2 standard).

Logical address
Reference to a particular storage location in which the user program executed in the programmable controller can read or write an input or output signal.

Logical base address

Logical address of the first input or output signal of a peripheral hardware module.

Loop resistance

Total resistance of the forward and return conductor of a bus cable.

LSAP

Link Service Access Point. Access point (address) of Layer 2.

Manchester code

A coding method used for sending digital data in a fieldbus system. The Manchester code transmits data and non-data information such as the clock pulse in a single, self-synchronizing signal (in accordance with IEC 1158-2, PROFIBUS PA).

Mandatory Services

Services which must be supported by every station connected to a PROFIBUS network.

Master/slave principle

Bus access method in a network in which one device, called the master, controls one or more devices, called the slaves. Only one station can assume the role of the master.

Memory bit

A 1-bit storage location. Basic STEP 7 operations can be used to read and write the contents of memory bits, memory bytes and memory words. The memory area can be used by the user program to store intermediate results.

Module parameters

Values that define the behavior and response of a module or a device connected to the PROFIBUS network. Depending on the module used, some of these parameters can be modified by the user program. Also called "dynamic data records".

MPI

*M*ulti *P*oint *I*nterface. Programming interface of SIMATIC S7 devices. It allows the simultaneous connection of several programming devices, text display devices, operator panels to one or several CPUs.

MPI address

Each module connected to an MPI subnet must be given its own MPI address.

Non-floating

On modules with non-floating inputs and outputs, the reference potentials of the control and load circuits are electrically connected to each other.

OB priority

The operating system of the CPU executes the different parts of the user program with different priorities. For example, the cyclical user program and a hardware-interrupt program belong to different priority classes. Each priority class is associated with specific organization blocks (OBs) in which the user programs the required reaction to an interrupt or event. The OBs have different priorities. Should several different types of interrupts occur simultaneously, then the higher priority OB interrupts a lower-priority OB. The priorities are set to default values, but they can be changed by the user.

Offline

A PG programming device or PC is in the offline state when it is not connected to a programmable controller for the purpose of exchanging information.

Online

A PG programming device or PC is in the online state when it is connected to a programmable controller for the purpose of exchanging information.

Operating mode

The programmable logic controllers (PLCs) of the SIMATIC S7 series can be in one of four different operating states: STOP, STARTUP, HOLD and RUN.

Operating mode HOLD

see HOLD, operating mode

Operating mode RUN

see RUN, operating mode

Operating mode STARTUP

see STARTUP, operating mode

Operating mode STOP

see STOP, operating mode

Operating system

The software that controls the allocation and usage of hardware resources such as memory, CPU time, disk space and peripheral devices. The operating system contains all the functions required to control and monitor the execution of the user programs, the allocation of the resources to the individual user programs and the operating mode of the hardware. The operating system is the foundation on which applications are built. Example: Windows

Operating system (CPU)

The operating system of the CPU controls and monitors all functions and program sequences that are directly related to the CPU, and not to control tasks in the plant.

Operator control and process monitoring systems

Systems that represent process information on a PC screen or operator panel display – often in graphical form –, and also allow the operator to enter commands that control process components.

Optically isolated

On modules with optically isolated inputs and outputs, the reference potentials of the load and control circuits are galvanically isolated from each other, for example by means of optocouplers, relay contacts, or transmitters. Input and output circuits can be rooted. Not to be confused with "floating".

Organization block

Software interface between the operating system of the CPU and the user program. The program code contained in the organization blocks determine the order of execution of the various parts of the user program.

Parameter

1. Variable of a STEP 7 program code block (see Block parameter, Actual parameter, Formal parameter). 2. Variable that sets the characteristics and response of a hardware module. A module have one or more parameters. When delivered by the manufacturer, a hardware module is usually set to meaning-

ful default parameters which can be modified using the STEP 7 program. We distinguish static and dynamic parameters.

Parameter, dynamic

Unlike static parameters, dynamic module parameters can be modified during online operation by means of system function calls (SFCs) in the user program. For example, the limit values of an analog input module are usually dynamic parameters.

Parameter master

Each DP slave device is associated with a so-called parameter master device which is a Class 1 master. In the startup phase, this master has the task of transmitting the parameter set to the DP slave. It has read and write access to the DP slave.

Parameter, static

Unlike dynamic parameters, static module parameters cannot be changed by the user program. You have to use STEP 7 to modify static parameters such as the input delay of a digital input module.

Parameters, assigning of

You define the response of a hardware module by setting and assigning parameters to a it.

PCMCIA

Personal Computer Memory Card International Association. An association of approximately 450 companies of the computer branch whose main goal it is to define international standards for the miniaturization and flexible usage of PC expansion cards. The PCMCIA standard defines the basic technique for the use of PC Cards on the computer market. The association cooperates with JEIDA (standard for PC cards for compact PC expansion modules).

Physical Layer

Transmission layer of a data communication network.In a PROFIBUS network, the transmission layer comprises a twisted-pair cable which acts as the transmission medium, termination resistors, connectors and bus interfaces.

Plant

also called "Installation"

The whole of the electrical resources making up a usually complex installation in manufacturing or process engineering industries. The industrial plant usually comprises, amongst other things, programmable logic controllers (PLCs), devices for operator control and process monitoring, bus systems, field devices, drives, supply cables.

PLC

Programmable logic controller. Electronic control device whose function is determined by the user program stored in its CPU. The installation and wiring of the plant to be controlled does not depend on the function of the PLC user program. This means that modifications in the logic interaction of the plant machinery is easily implemented by reprogramming the PLC. The PLC can be considered as an industrial PC. It comprises a CPU with a memory, input and output modules and an internal bus system. The peripheral devices and the programming language are designed to meet the requirements of process and plant control in industrial applications.

Power supply, external

Power supply for I/O modules.

Priority

In SIMATIC S7 systems, you assign priorities to organization blocks and thus define which organization block has the right to interrupt the cyclical user program, and under which conditions. Higher-priority organization blocks (OBs) may interrupt lower-priority organization blocks (OBs).

Priority classes

The operating system of a SIMATIC S7 CPU provides up to 28 priority classes to which you can assign organization blocks (OBs) for interrupt-driven processing. The priority classes define which OB has the right to interrupt another OB, and at which conditions. If one priority class contains several OBs and these are called simultaneously due to interrupts, then these OBs of equal priority do not interrupt each other; they are processed sequentially.

Process image (table)

Special memory area in the CPU in which the signal states of the digital input and output modules are stored. We distinguish between the process-image output table (PIQ) and the process-image input table (PII).

Process interrupt

also called "Hardware interrupt"

Modules with interrupt capability can generate a process interrupt as a reaction to a certain event in the process. The process interrupt is transmitted to the CPU. According to the priority of the interrupt, an organization block (OB40 to OB47) is called and processed. This OB contains a user program which describes the reaction to the event or fault that caused the interrupt.

Process-image input table (PII)

The operating system reads the signal states of the input modules prior to each new cycle of the user program and stores this information in the process-image input table.

Process-image output table (PIQ)

The operating system transfers the process-image output table to the output modules at the end of each cycle of the user program.

PROFIBUS

*PRO*cess *FI*eld *BUS*; European standard for process and fieldbus systems, specified in the PROFIBUS EN 50 170 standard, Volume 2, PROFIBUS. PROFIBUS defines the functional, electrical and mechanical characteristics of a bit-serial fieldbus system. PROFIBUS is a data communication network which interconnects PROFIBUS-compatible automation systems and field devices on the cell and field levels of an automated plant. PROFIBUS networks can use the communications protocols "DP" (= Distributed Peripherals), "FMS" (= Fieldbus Message Specification), and "PA" (Process Automation).

PROFIBUS address

Each station (also called "node") connected to a PROFIBUS network must be identified unambiguously by its PROFIBUS address. PG programming devices and PCs connected to a PROFIBUS network have the default address "0". The other stations on the network may be given addresses in the range from 1 to 125.

PROFIBUS DP

Acronym for "Process Field Bus for distributed peripherals." Standardized specification (EN 50170) for an open-architecture fieldbus system. PROFIBUS DP is primarily used for time critical applications in automated manufacturing industries.

PROFIBUS FMS

Acronym for "Process Field Bus using the FMS protocol". Standardized specification (EN 50170) for an open-architecture fieldbus system. PROFIBUS FMS is primarily used for the exchange of process information in automated manufacturing.

PROFIBUS PA

Acronym for "Process Field Bus for Process Automation". Standardized specification (EN 50 170) for an open-architecture fieldbus system. PROFIBUS PA is primarily used for the exchange of process information in automated process control systems.

Program execution, event-controlled

With event-controlled program execution, the user program is interrupted by start events according to defined priorities. When such a start event occurs, the software block in progress is interrupted after completion of the currently processed statement, and the organization block (OB) associated with the event is called and processed. After completion of this OB, cyclical program execution continues at the point of interrupt.

Programmable logic controller

Programmable device that processes input and output information according to a user program, in order to automate a technological process or plant. A PLC is not an autonomous system; it should always be seen in connection with the process or plant to be controlled.

Project

An S7 project contains all objects that make up a specific automation task, including all stations, hardware modules, and their interconnection in a network.

Protocol

see Communications protocol

Protocol Data Unit

A PDU (Protocol Data Unit) is a data packet that contains the information to be transmitted from one station to another in a network.

Rack

A mounting rack is made up of plug-in slots for the insertion of hardware modules (also called "cards").

Reaction to errors

Reaction to a runtime error. The operating system can react in different ways: the PLC is switched to the STOP state, an error organization block is called, or the error is displayed. Note that the error OB contains a user program with instructions on how to react to the error.

Repeater

Connects one network segment with another and regenerates the signals to be transmitted.

Response monitoring

If a DP slave is not addressed within the set response monitoring time, then it goes into a so-called "safe" state. This means that the DP slave sets its outputs to "0". During system configuration, you can enable or disable response monitoring separately for every single DP slave device.

Reference potential

Potential which serves as the point of reference from which all voltages of the associated circuits can be viewed or measured.

RUN, operating mode

When the CPU is in the RUN mode, it executes the user program and updates the process image at regular intervals (i.e., cyclically). In this mode, all digital outputs are enabled.

S7 program

The S7 program is executed in the CPU of a SIMATIC S7 programmable controller. It contains software blocks, data sources and logic operations to control the S7 modules that are plugged into the SIMATIC S7 rack.

S7 protocol

The S7 protocol (also called "S7 communication" or "S7 functions") is a simple and efficient interface between SIMATIC S7 stations and the PG/PC programming device and HMI systems.

SCADA system

Supervisory control and data acquisition system. Today, SCADA systems are usually PC-based. They collect process data from the PLC, visualize this information in graphical form and act as a human machine interface for the plant operator. This means a SCADA system allows the operator to enter commands in order to control plant components.

Segment

The bus line between two terminating resistors. One bus segment may contain up to 32 stations (also called "nodes"). Segments are connected with each other by inserting RS 485 repeaters between two adjacent segments.

Services

Services (methods of exchanging data) offered by a communications protocol.

SFB

An SFB (*SystemFunctionBlock*) is a function block integrated into the operating system of the CPU that can be called in the STEP 7 user program if required.

SFC

SystemFunctionCall. A function that is integrated in the operating system of the SIMATIC S7 CPU. An SFC can be called by the STEP 7 user program.

Shield impedance

A.C. resistance of the cable shield. The shield impedance is one of the characteristic values of a cable and is usually specified by the manufacturer of the cable.

Short-circuit

Reduction of a potential difference between two points in a circuit to zero by connections of a conductor of zero impedance, in which no power is dissipated. If the short-circuit is not intended, damage may result if the circuit is not opened quickly elsewhere.

SIMATIC Manager

Graphical user interface for SIMATIC applications for Windows 95 and Windows NT. SIMATIC Managers provides all the functions and tools required for configuring a SIMATIC S7 system and defining its parameters.

Slave

Device that is controlled by another, the master. In a data communications network, a slave may only exchange data with the master upon request by the master. In a PROFIBUS DP network, SIMATIC devices such as ET 200B, ET 200C, etc. are used as slaves.

Start event information

Part of an organization block (OB). The start event information indicates the event that has invoked the call of the OB. The start event information contains: the event ID (event class, event IDs, event number), an even time stamp, and some additional information, such as the address of the signal module that has generated the interrupt.

STARTUP

Transitional operating mode in which the PLC switches from the STOP state to the RUN state.

Startup OB

Depending on the position of the startup selector switch (applies to S7-400 only) and the cause for the restart (recovery after supply failure, changeover from STOP to RUN using the mode selector switch or the PG programming unit), a startup organization block is called in the user program. The startup OB can generate a "warm restart" or a "hot restart" (applies to S7-400 only). The program contained in the startup OB is written by the SIMATIC S7 user and is often used to set default values that ensure a safe startup of the plant after power failure.

STARTUP, operating mode

STARTUP is the transition from the operating mode STOP to RUN. The PLC can be started up by using the mode selector switch on the PLC, by switching on its power supply, or by means of a command on the programming device. A PLC of the S7-300 performs a "warm restart". A PLC of the S7-400 series, performs a "warm restart" or a "hot restart," depending on the position of the startup selector switch on the PLC.

Statement
A STEP 7 statement is the smallest self-contained unit of the STEP 7 user program. It is created using a textual programming language. The statement contains a command addressed to a process component.

Statement list
A form of representation of the STEP 7 user program. The statement list can be considered as the assembler programming language of STEP 7. A user program programmed in STL is made up of statements each of which is a program step that is executed by the CPU.

Station address
Unambiguous identification by which a device (e.g., PG) or a programmable module (e.g., CPU) is known and addresses in a subnet (e.g., MPI, PROFIBUS).

Status interrupt
DPV1 slaves can report detected status changes to the CPU via status interrupts.

STEP 7
Programming software for the creation of user programs for SIMATIC S7 applications.

STEP 7 programming language
Programming language for programmable controllers of the SIMATIC S7 series. STEP 7 allows programming in three different forms of representation: statement list STL, function block diagram FBD, ladder logic LAD.

STL
→ Statement list

STOP, operating mode
There are three events that can cause the CPU to go to the STOP state: switching the PLC mode selector switch to STOP; occurrence of a CPU fault; STOP request transmitted by PG programming device. When the CPU is in the STOP state, it does not execute the user program. All modules are switched to a safe state. Operator control and process monitoring functions and some programming functions can still be carried out in the STOP mode.

Subnet
Entirety of all physical components that make up a data transmission path and of the procedures that govern the exchange of information on that path. The stations (also called "nodes") that belong to the subnet are connected with each other without network crossover. Examples of subnets: MPI, PROFIBUS, Industrial Ethernet.

SYNC, control command
Control command sent by the DP slave to a group of DP slaves. Upon the arrival of a SYNC command, the DP slaves maintain the states of their outputs stationary. When other telegrams arrive, the DP slave saves the output data, but keeps the states of the outputs unchanged. With each newly arriving SYNC command, the DP slave sets the outputs to the output values it has received and stored. Cyclical updating of the output states only resumes when the DP master has canceled the SYNC mode by sending an UNSYNC command.

System data block (SDB)
Special data areas in the CPU that contain system settings and module parameters. System data blocks are generated and modified during configuration of an S7 project.

System diagnostics
Detection and evaluation of system error information.

System function SFC
see SFC.

Ttr
Time target rotation. In a master/slave communications network, every master compares the time target rotation with the actual time required for the token to circulate. The difference defines the time available to the master to transmit its own data telegrams.

Termination resistor
Resistor connected to the ends of a cable to prevent a reflection at these points of line

discontinuity. With PROFIBUS, the cable ends or segment ends must always be terminated by means of a termination resistance.

Token

A unique structured data object or message that circulates continuously among the nodes of a token ring. It describes the current state of the network and determines which station is allowed to transmit. Only the active station (master station) that holds the token can transmit information to other active and passive stations. When the data cycle is completed, the token is passed on to the next active station.

Token ring

A master station connected to a network receives the token and holds it for a brief time, in which it marks the token as being in use and transmits information through the network. It then releases the token and passes it to the next master station. The master stations are connected to a token ring network.

Token rotation time

The time it takes for the token to circulate the entire network. In other words, the time that elapses from the moment a station has received the token to the next time the same station receives the token again.

Tool

Software used for designing, programming and configuring an automation project.

UNFREEZE

→ FREEZE

UNSYNC

→ SYNC

Update interrupt

DPV1 slaves can report internal updates to the CPU using an update interrupt.

Uploading to the programming device

Copying objects of a configuration such as blocks of program code from the load memory of the CPU to the programming device, either directly to the programming de-

vice that is connected to the CPU, or indirectly to a remote programming device connected to the PROFIBUS network.

User program

Contains all the statements, declarations and data for the signal processing that is required to control an industrial process or plant. The user program is structured into smaller program units, so-called blocks (blocks), and is assigned to a programmable module (Module, programmable), such as a CPU or FM.

Variable

Data item whose contents is variable. It can be addressed by the STEP 7 user program. A variable is made up of the address (e.g., M 3.1) and the data type (e.g., Boolean) and may be given a symbolic name (e.g., Motor_-ON).

Variable declaration

Defines the characteristics of a variable. Declaring the variable consists in giving it a symbolic name, defining its data type and address, and a default value and comment, if required.

Vendor-specific interrupt

DPV1 slaves can report vendor-specific events to the CPU via manufacturer interrupts.

VFD

Virtual Field Device. Producing an image of a physical field device helps to create a uniform and coherent view of field devices.

Warm restart

When the CPU is started by moving its mode selector switch from STOP to RUN, or by switching on its power supply, the startup organization block OB 101 (hot restart, for S7-400 systems only)) or OB 100 (warm restart) is executed before cyclical program execution starts. The warm restart procedure first reads the process image of the inputs and then executes the first statement of the cyclical organization block OB1.

Abbreviations

A

AI   Analog input
AO   Analog output (STEP 7)
ASI   Actuator-sensor interface
AWG   American Wire Gauge

C

CP   Communications Processor
CPU   Central Processor Unit

D

DB   Data Block
DIN   Deutsche Industrie Norm (German Industrial Standards)
DP   Decentralized Peripherals, PROFIBUS protocol
DS   Data Set
DSAP   Destination Service Access Point

E

ED   End Delimiter
EIA   Electronic Industries Association, USA
EMC   ElectroMagnetic Compatibility
EN   Europäische Norm (European Standards)
EN 50 170   European standard for PROFIBUS DP and FMS.
               Succeeds the German DIN 19245 standard.

F

FBD   Function Block Diagram (STEP7)
FC   Frame Control, telegram control byte
FCS   Frame Check Sequence, telegram check byte
FDL   Fieldbus Data Link Layer (2), protocol layer 2 of PROFIBUS
FM   Function Module
FMS   Fieldbus Message Specification,
               application services of PROFIBUS Layer 7,
               FMS can be operated in conjunction with DP.

G

GAP   Among active bus stations, the address range (gap) from one station address to the next station address
GAPL   GAP List, List of all stations in the current GAP area

H

HD   Hamming Distance
HMI   Human Machine Interface
HSA   Highest Station Address

I

IEC   International Electrotechnical Commission
IEEE   The Institute of Electrical and Electronics Engineers, USA
IM   Interface Module
ISO/OSI   International Standards Organization/Open Systems Interconnection

L

LAD   Ladder Diagram (STEP7)
LAS   List of Active Stations
LE   Length, length byte
Ler   Length repeated, length byte repeated
LLI   Lower Layer Interface, part of the Application Layer (7) for PROFIBUS FMS (Link to Layer 2)
LSAP   Link Service Access point, service access point to Layer 2
LSB   Least Significant Bit

M

MAC   Medium Access Control; it determines when a device is given the right to send data
MPI   MultiPoint Interface, standard interface of the SIMATIC S7 devices

N

NRZ   Non-Return-To-Zero

O

OB   Organization Block
OLM   Optical Link Module
OLP   Optical Link Plug
OP   Operator Panel
OSI   Open System Interconnect

P

PA	Process Automation, PROFIBUS definition for process automation according to IEC 1158-2 and DIN E 19245, Part 4.
PG	Programming device
PI	PROFIBUS International
PLC	Programmable Logic Controller
PNO	PROFIBUS User Organization
PS	Power Supply

S

SA	Source Address, source address byte
SAE	Source Address Extension
SAP	Service Access Point, unambiguous identification of the data to be sent or requested within a telegram. Every telegram contains a source SAP and a destination SAP (exception: Data is exchanged through the Default SAP).
SC	Single Character, individual character (short acknowledgment) with PROFIBUS
SD	Start Delimiter
SDA	Send Data with Acknowledge
SDB	System Data Block
SDN	Send Data with No Acknowledge
SFB	*System*Function*Block*
SFC	*System*Function*Call*
SRD	Send and Request Data
SSAP	Source Service Access Point
SSL	System Status List
STL	StaTement List

T

tbit, Tbit	Time unit for the transmission of one bit on PROFIBUS (reciprocal value of the transmission rate, Example: for 12 Mbaud, 12,000,000 bits/sec, the tbit value is 83 nsec).

V

VAT	Variable table (STEP7)
VDE	Verein Deutscher Elektroingenieure (German Association of Electrical and Electronics Engineers)
VFD	Virtual Field Device. That part of a real device which is accessible for data communication.

Standards and Regulations

[1] EN 50170 Volume 2
General Purpose Field Communication System
Volume 2: Physical Layer Specification and Service Definition

[2] DIN E 19245
German Fieldbus Standard – PROFIBUS (Draft)

[3] IEC 870-5-1
Telecontrol equipment and systems, Part 5: Transmission protocols;
Section 1: Transmission frame formats

[4] EN 60870-5-1
Telecontrol equipment and systems, part 5: transmission formats,
Main section: Part 1 Telegrams

[5] DIN 19241
Control and Instrumentation: Bit-serial process buss interface system;
Elements of the transmission protocol and message structure

[6] IEC 955
Process data highway, type C (PROWAY C), for distributed process
control systems

[7] IEC 1158-2
Fieldbus standard for use in industrial control systems – Part 2: Physical
layer specification and service definitions

[8] EIA RS 485
Standard for electrical characteristics of generators and receivers for
use in balanced digital multipoint systems

[9] PNO Guidelines
Optical Transmission Techniques for PROFIBUS
Version 1.1 from 07.1993

[10] IEC 61158
Globally binding standard for PROFIBUS and other fieldbuses. PROFIBUS DP is
defined in Part 3 of this standard.

Index